Earth Science Investigations

Edited by

MARGARET A. OOSTERMAN

and

MARK T. SCHMIDT

SERVING THE GEOLOGICAL SCIENCES AGI 1948

American Geological Institute
Alexandria, Virginia

The cover illustration was adapted from a colorful map that the U.S. Geological Survey and the Smithsonian Institution produced. The large (3 ft. 5 in. x 4 ft. 10 in.) map shows the Earth's physiographic features overlain by its volcanoes, earthquake epicenters, and the movement of its major tectonic plates. Prominent global features are visible from a distance on the 1:30 000 000-scale map, and a closer inspection reveals more detailed features.

This inexpensive map is for sale from the U.S. Geological Survey, Map Distribution, Box 25286, Federal Center, Denver, CO 80225 and from the National Museum of Natural History, Smithsonian Institution, Washington, D.C. 20560. Ask for "This Dynamic Planet, World Map of Volcanoes, Earthquakes, and Plate Tectonics," compiled by Tom Simkin, Robert I. Tilling, James N. Taggart, William J. Jones, and Henry Spall, 1989.

Given the chance, many students would love to look inside a volcano. The U.S. Geological Survey and the Smithsonian Institution do just that in the educational film "Inside Hawaiian Volcanoes." The teacher's guide, which contains questions and lab exercises, is U.S. Geological Survey Open-file Report 89-685. Contact any of the Earth Science Information Centers listed inside the front cover of this manual for ordering information.

The film is available in VHS video cassette and 16-mm motion-picture formats. The Smithsonian Institution distributes the cassettes from the Department of Mineral Sciences, NHB 119, Smithsonian Institution, Washington, D.C. 20560. For information on the 16-mm films, contact the U.S. Geological Survey, Visual Information Services, 790 National Center, Reston, VA 22092, or phone (703) 648-4379.

Director of Education: *Andrew J. Verdon, Jr.*
Director of Publications: *Julia A. Jackson*
Project Coordinator and Editor: *Margaret A. Oosterman*
Editor and Page Compositor: *Mark T. Schmidt*
Book Designer: *Melody Oakes*
Illustrator: *Julie De Atley*
Clerical Support: *Norma Adams and Laura D'Sa*
Field-testing Coordinator: *Sue Cox*

Composed in Century Old Style with Quark XPress 2.12 on a Macintosh SE
Printed by United Book Press, Inc., on 60-lb. James River, a recycled, acid-free paper

recycled paper

Printed and bound in the United States of America

International Standard Book Number: 0-922152-07-1

Contents

Preface

The plan for this book evolved from conversations with earth science teachers, who voiced the need for innovative laboratory exercises. They stressed the importance of student materials that involve a "hands on—minds on" approach. They sought activities that teachers have specifically developed for and successfully used with secondary-level students. We produced *Earth Science Investigations* in the framework of suggestions from many earth science teachers. We hope that the investigations will challenge and motivate your students.

The 27 exercises are grouped by general subject, but the grouping does not imply a specific order. Thus, a teacher can select and reproduce for students the exercises that fit best with his or her curriculum and resources. Our aim is to provide accurate, ready-to-use activities that teachers can integrate with any earth science text or program. All but one of the exercises are clearly for students. The exception, "Instructions for Assembling Astronomy Materials" by Paul Johnston (p. 211), has been written for teachers. It could easily be adapted, however, as a student activity. Each of the earth science activities has been tested with students and reviewed for scientific accuracy. Because classes and classroom setups vary, teachers may wish to modify an investigation to suit their situation. Be sure to do each exercise before having students do it.

The investigations are intended to be done under the direct supervision of a qualified earth science teacher. Cautionary notes are provided to alert teachers and students to proceed with caution and to use safe procedures. General cautions appear near the beginning of a lab. Specific cautionary notes appear at the point in a lab where they apply. Some of the exercises involve chemicals that may be harmful if they are misused, or if the procedures described are not followed. Chemicals should be used in a well-ventilated classroom, and protection for eyes and clothing should be worn. The American Geological Institute, the authors, and the editors are not responsible for any accidents or injuries that may occur during or as the result of performing any of the exercises in this book.

Each investigation begins with a section of clearly stated **Objectives** for students. **Time required** for doing an activity is given in class periods with the assumption that one class period runs 45-50 minutes. The amount of time given does not include pre-lab and post-lab instruction and discussion or optional sections of labs.

The **Materials** section lists materials needed other than the lab itself and pencils. If no additional materials are needed, that is noted. Materials are grouped under subheadings such as *per student* or *per group of 2-4 students.* Throughout the labs, metric units have been used where appropriate, and approximate English conversions follow in parentheses.

The **Background** section contains information and vocabulary that will help students work through each lab. We have tried to keep this section brief and understandable without oversimplifying the scientific content. The information provided is not intended as a comprehensive presentation on a particular subject. Most of the labs include an **Additional resources** section listing references on the subject.

The **Procedure** section outlines exactly how a student or group of students should proceed. In the **Questions** section, the editors have purposely omitted room for answers to avoid the possibility of students writing to fill an arbitrary space. Generally, students will find the information they need to complete the worksheets and answer the questions in the lab itself. Answer keys have been provided as needed.

If you have questions or comments about any of the investigations, please contact the editors at the National Center for Earth Science Education, American Geological Institute, 4220 King Street, Alexandria, VA 22302-1507.

Acknowledgements

Earth Science Investigations is the result of a cooperative effort by many professionals, who generously gave their expertise and time to the project. Initially, we invited members of the National Association of Geology Teachers, National Earth Science Teachers Association, and National Science Teachers Association to submit their best, original, classroom-tested exercises.

We are grateful to the teachers who sent exercises for consideration, to the teacher consultants and panels who evaluated the activities, to the earth scientists who reviewed them, and to the field-test teachers and students who performed the investigations. Our special thanks go to the following individuals:

Teacher consultants: Jeffrey C. Callister, Newburgh Free Academy; Sue Cox, U.S. Geological Survey; Jane Crowder, Issauga Middle School; and Richard H. Lefebvre, Grand Valley State University

Teacher review panels: Jeanne Endrikat, Nancy Roach, and James Sproull, Fairfax County Public Schools; Jeannie Allison, Richard S. Smith, Larry Weatherwax, and their science supervisor, Emma Walton, Anchorage School District

Scientific reviewers: Julius Dasch, NASA Headquarters, Washington, D.C.; Ira W. Geer, State University of New York, Brockport; John E. Moore, Environmental Strategies Corporation; Orrin Pilkey, Duke University; Bernard W. Pipkin, University of Southern California; Tom Simkin, Smithsonian Institution; Peter H. Schultz, Brown University; and Dale A. Springer, Bloomsburg University

Field-test teachers: Terri Aquino, Ron Armstrong, Pamela Bartkowski, Arnold Becker, Brigitte Berry, Donald Birdd, Carolyn Bongi, David Cardwell, Robert Christman, David Crawford, Jane Crowder, Patricia Davenport, Donato DiGregorio, Donald Ehrenberger, Jeanne Endrikat, Russ Fazio, Jack Fearing, Theron Garcia, Richard Hamilton, Jacob Hirschfelt, Charles Hofbauer, Kathy Jacquez, Joyce Johnson, William Johnson, Paul Johnston, Sally Jones, William Kent, Kenneth Knezic, Linda Knight, Robert Lamons, Priscilla Lee, Mary Alice Legendre-Cain, Michael McCauley, Robert McDowell, Thomas McGuire, Pete Matthews, George Moon, Thomas Morganti, JoAnn Mulvany, Mark Ness, Bernard Noeller, Jim Palm, Alfred Palmer, Sister Mary Parrott, Randy Patrick, William Philips, Maryanne Porter, John Reid, Frank Roberts, Susan Schmitt, Joe Schrock, Treasure Sexton, Patty Shinstrom, Vicki Simon, James Sproull, Richard Staley, Joy Stouffer, Bernadette Tomaselli, Laurie Trippett, Ken Uslabar, Christopher Visco, Patricia Warthan, Marilyn Watts, Wally White, Samuel Wine, and Frank Zuerner

Andrew J. Verdon, Jr.
Director of Education

Eolian Action

Maryanne E. Porter
Academy Park High School
Sharon Hill, Pennsylvania

Objective

To observe and record movement and deposition caused by the wind by constructing a geologic model that shows the formation of different types of sand dunes.

Time required

One class period; however, a setup may be kept running for an entire school day to measure the long-term effects of constant wind.

Materials

per class

Stream table: 1.25 m x 30 cm x 5 cm (4 ft. x 1 ft. x 2 in.). Any type of plastic, wooden, or metal tray can be used.
Fan: adjustable speeds if possible
Play sand: enough to fill the stream table almost to the top
Small bag of colored sand
Meter stick
3 plexiglass sheets: two should be the length of the stream table, one the width. The height should be about one foot. The sheets keep the sand from blowing out of the stream table, but they do not block the student's view. Any type of transparent barrier can be used.
Washable marker

Background

In Greek mythology, Aeolus was god of the wind. Aeolian (or eolian) action refers to action by the wind. Sand dunes are mounds, ridges, or hills of wind-blown sand. Wind or eolian action forms dunes by moving and depositing sand particles.

Wind carries sand from the **windward** side of a dune and deposits it on the **leeward** side. **Sand dune migration** is the movement of a dune by the wind over a long period of time. The direction of the migration follows the direction of the wind, that is, from windward to leeward (Figure 1).

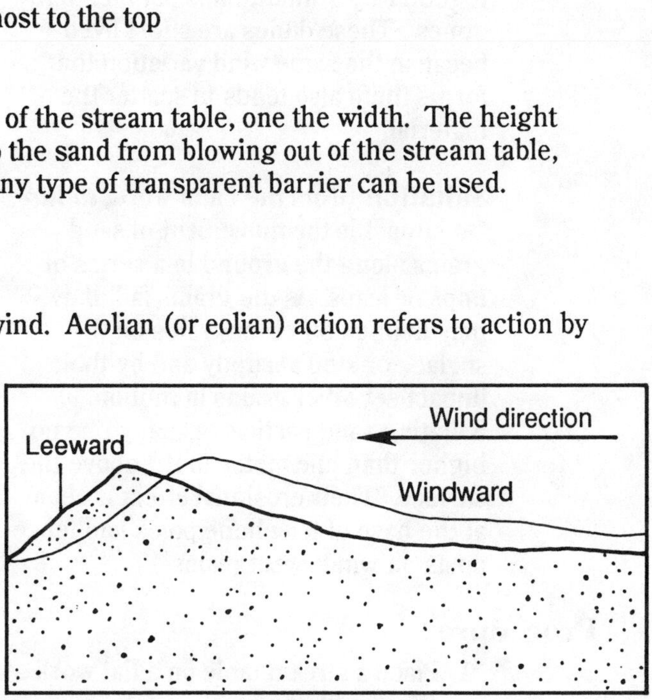

Figure 1. Sand dune migration.

There are four major types of sand dunes (Figure 2). **Barchans**, or crescent dunes, develop in flat desert areas where wind direction and speed are fairly constant. These dunes often form in clusters and are crescent shaped with the horns pointing in the leeward direction.

Parabolic dunes develop in areas with large amounts of sand. They are described as spoon shaped because of the deep depressions that develop inside the horns. The horns of a parabolic dune point in a windward direction. Vegetation often anchors the horns.

Transverse dunes develop along shorelines. They are a series of dunes that are parallel to each other and the water's edge. Dune crests tend to be at right angles to the direction of the wind.

Longitudinal dunes are the result of great numbers of barchans being blown together by a wind that is parallel to the dunes . These dunes are short lived because the same wind variation that forms them also tends to scatter the material.

Saltation (from the Latin word, *saltare,* "to jump") is the movement of sand grains along the ground in a series of hops or leaps. As the grains fall, they may bounce off other grains on the surface or stop abruptly and by their impact set other grains in motion. Saltating sand particles usually rise no higher than one meter (3 ft.) above the surface. Their erosional effect is seen at the base of telephone poles and fence posts on wind-swept plains.

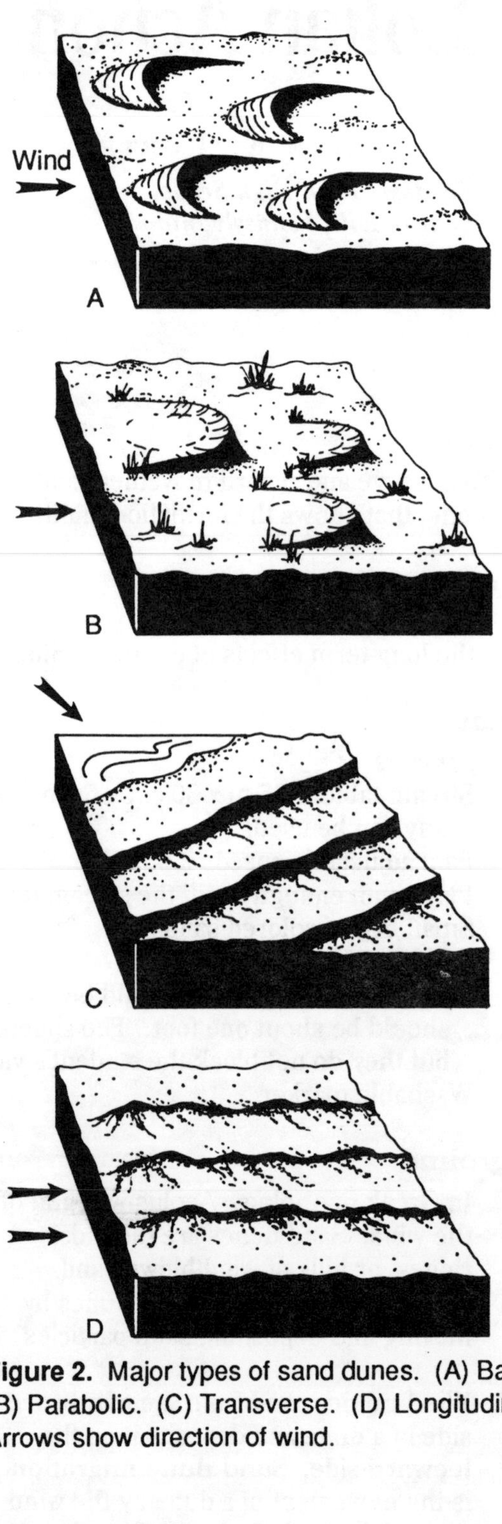

Figure 2. Major types of sand dunes. (A) Barchan. (B) Parabolic. (C) Transverse. (D) Longitudinal. Arrows show direction of wind.

Procedure

1. Place a stream table on a flat work surface, and fill it almost to the top with play sand (Figure 3).

2. Stand two long plexiglass sheets vertically in the sand along the sides of the stream table, and stand a third sheet at one end.

3. Position a meter stick horizontally along one length of the stream table.

4. About 20 cm (8 in.) from the open end of the stream table, pour a line of colored sand on top of the play sand, across the width of the stream table.

5. Place a fan at the open end of the stream table, and turn it on. **CAUTION:** *Be careful of the fan's blades, and stay away from blowing sand particles.*

   You will notice changes in the position of the sand immediately. The longer you can observe the activity, the more changes you will see.

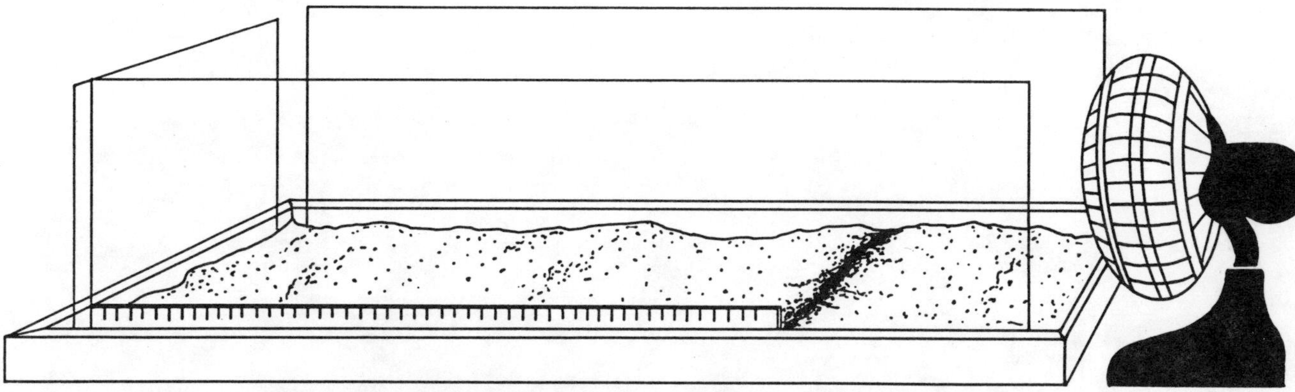

Figure 3. Setup diagram. Adapted from a photograph taken by Scott Wynn.

6. Depending on the types of formations to be observed, the speed of the fan will have to be adjusted. If the fan does not have adjustable speeds, change the fan's distance from the end of the stream table to control the speed of the air hitting the sand.

7. Use a marker on the plexiglass to show the locations of the dunes at various time intervals.

8. Using a meter stick, measure the position of the dunes.

9. Record the starting time and the time relationships to the distance moved.

10. Make drawings of all the formations you observed.

Questions

Write a lab report using the following guidelines:

1. What is the problem? (Hypothesis)
2. How did you study the problem? (Materials and methods)
3. What did you find? (Results)
4. What did your results mean? (Discussion from textbook(s) or from teacher-generated questions)

Beach Profile Investigation

Christopher Visco
Sachem High School South
Lake Ronkonkoma, New York

Objective

To study the effects of beach erosion and deposition by graphing and comparing two beach profiles.

Time required

Two class periods

Materials

per student
Glue
Colored pencil or marker
Construction paper: 8 1/2 in. x 11 in.
Scissors

Background

Note: Students should know how to plot data on a graph.

The shoreline is a dynamic environment, affected by many processes including waves, tides, wind, and human activities. It is important to understand the long-term trends of erosion and deposition on a beach, because the beach helps prevent damage to man-made structures. One method of analyzing a shoreline is to take profiles or cross sections of a beach over a period of time. Figure 1 shows a generalized beach profile. By comparing successive profiles, it can be determined whether waves are building up the beach, deposition, or taking material away, erosion.

Procedure

Fire Island is a barrier beach located south of Long Island, New York. It is typical of the barrier beach systems found along the eastern coast of the United States. A barrier island is a ridge of sand that is parallel to a shoreline and about three to 30 km (two to 19 mi.) offshore.

Two profiles of the beach were made at the same location over a period of six months. One was made during January (winter) and the other during June (summer). The data for each profile are found in Table 1.

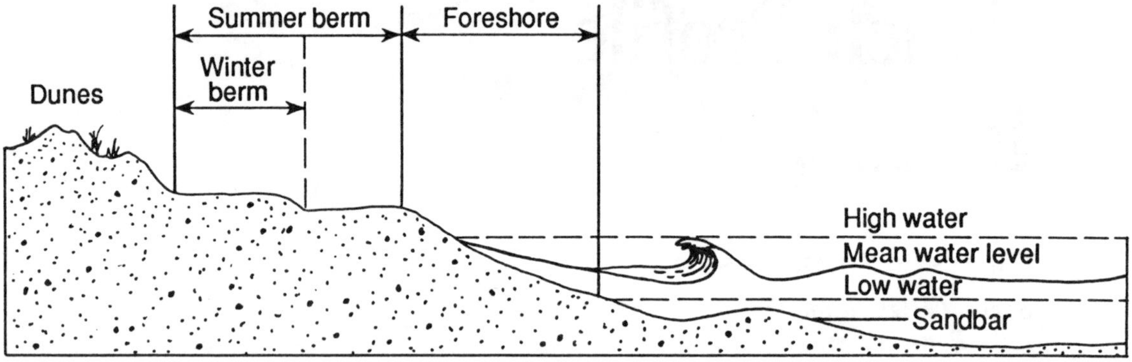

Figure 1. Generalized summer beach profile.

A beach profile shows several features that may change as a result of seasonal weather conditions. During the winter months, waves are generally high, and there are more storms, producing more frequent waves. Waves and tide action tend to erode the beach by carrying away material that may be deposited offshore, forming an underwater **sandbar**. Because of this erosion, the **berm**, the nearly horizontal section of a beach usually known as "the beach," is narrower in the winter months than in the summer months. The **foreshore**, the section between the low tide shoreline and the beginning of the berm, is less steep in winter than in summer. The foreshore is regularly covered and uncovered by the rise and fall of the tide.

During the summer months, waves are generally low and flat and tend to deposit sand and other particles on the beach. This deposition builds up the berm, making it wider, and steepens the foreshore. Sandbars may disappear.

1. Use a colored pencil or marker to mark sea level at a height of zero feet on the two graphs (Figure 2) on the worksheet.

2. Using Profile A data in Table 1, plot the height of the beach and the distance from the dunes on Profile A on the worksheet. Connect the points with a smooth line.

3. For Profile B data in Table 1, repeat step 2, on Profile B on the worksheet.

4. Cut along each profile line, and discard the portion above it.

5. Glue Profile A, the higher one, onto a sheet of construction paper. Position Profile B on top of Profile A so that the two profiles are aligned along the vertical and horizontal axes. Glue Profile B into place.

6. Label the following zones on each profile: dunes, berm, foreshore, and sandbar.

Table 1
Beach profiles of Fire Island, New York.*

Profile A		Profile B	
Height above sea level (ft.)	Distance from dunes (ft.)	Height above sea level (ft.)	Distance from dunes (ft.)
3.4	0	3.2	0
3.2	6	3.0	8
3.0	12	2.8	16
2.8	24	2.6	26
2.8	30	2.4	36
2.8	36	2.0	44
3.0	54	1.6	50
2.8	64	1.0	60
2.4	72	0.6	68
1.8	78	0.2	78
1.4	86	-0.2	88
0.6	98	-0.6	98
0.0	108	-1.0	110
-0.4	118	-1.0	118
-0.4	122	-0.8	128
-0.2	130	-1.4	142
-0.6	136		
-0.8	138		
-1.2	146		
-1.4	156		

*English units are used rather than metric units because the data were collected in English units. Conversion to metric units would result in fractional data that are more difficult to use.

Questions

1. Which profile was made in January? How can you tell?
2. Which profile was made in June? How can you tell?
3. Why do beaches typically show seasonal variations?

Beach Profile Investigation

Worksheet

Name ....
Class ....
Date ....

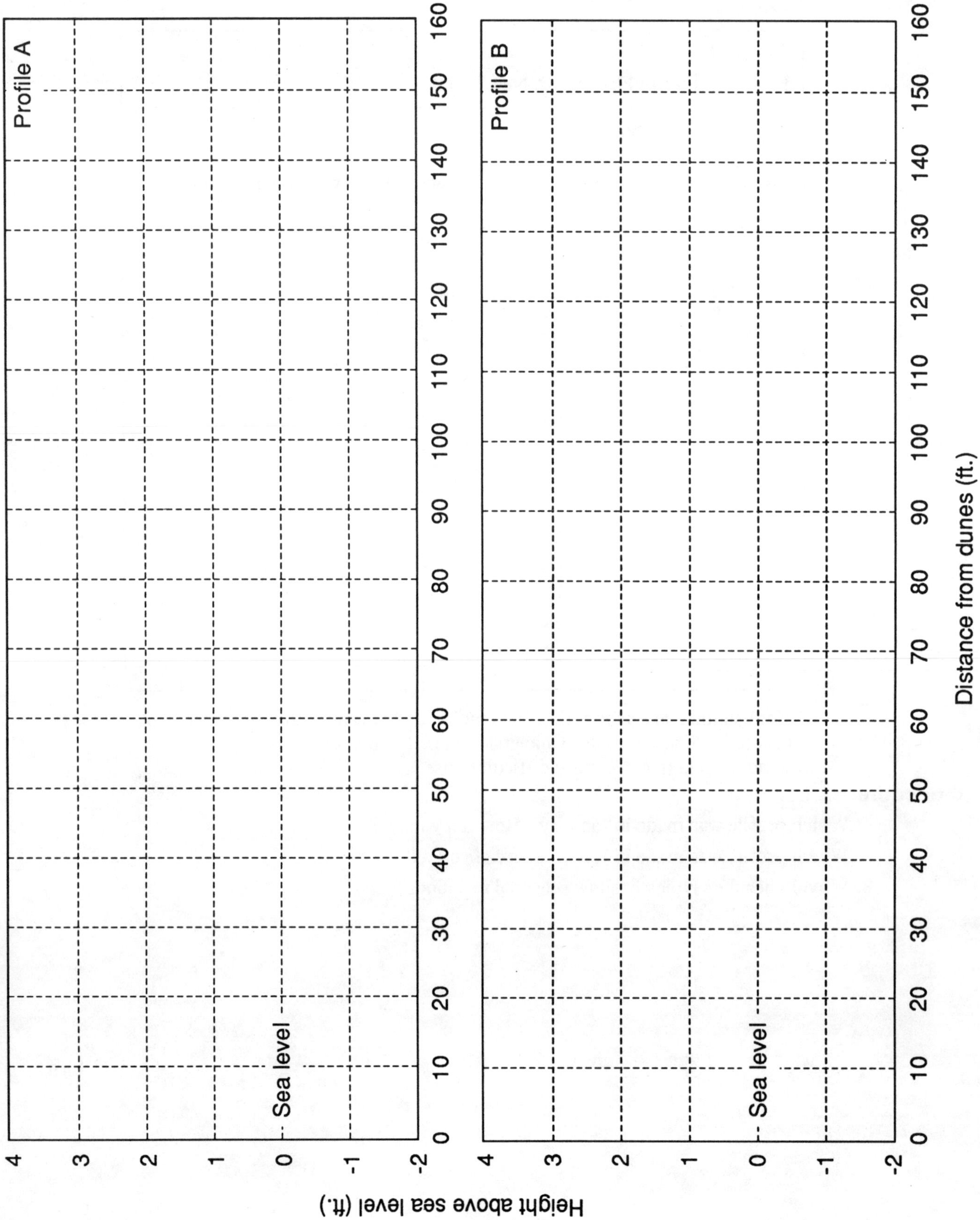

Figure 2. Summer and winter beach profiles of Fire Island, New York.

Answer ✓ Key

Procedure

See the next page for the worksheet answer key.

Questions

1. Profile B was made in January. It is lower in elevation, and the berm is narrower. Although the winter profile is lower, the winter berm is higher than the summer berm. An unusual number of spring storms occurred before the data were collected. Such activity probably accounts for the unusual height of the submerged sandbar, and it demonstrates that a higher sandbar does not only form in winter.

2. Profile A was made in June. It is higher in elevation, and the berm is wider. The summer profile is higher; however, the summer berm is lower than the winter berm. No data were collected beyond 142 feet because of the cold water temperature and the lack of a wet suit. It is quite likely that another submerged sandbar, higher than the one in the June profile, was present farther out from the coast.

3. The seasonal variation in a beach is a result of the different weather patterns associated with summer and winter. During winter, there are more storms, waves are larger and more frequent, thus erosion is dominant. During summer, storms are infrequent, waves are smaller, and offshore sand is nudged back towards the beach.

Answer ✓ Key

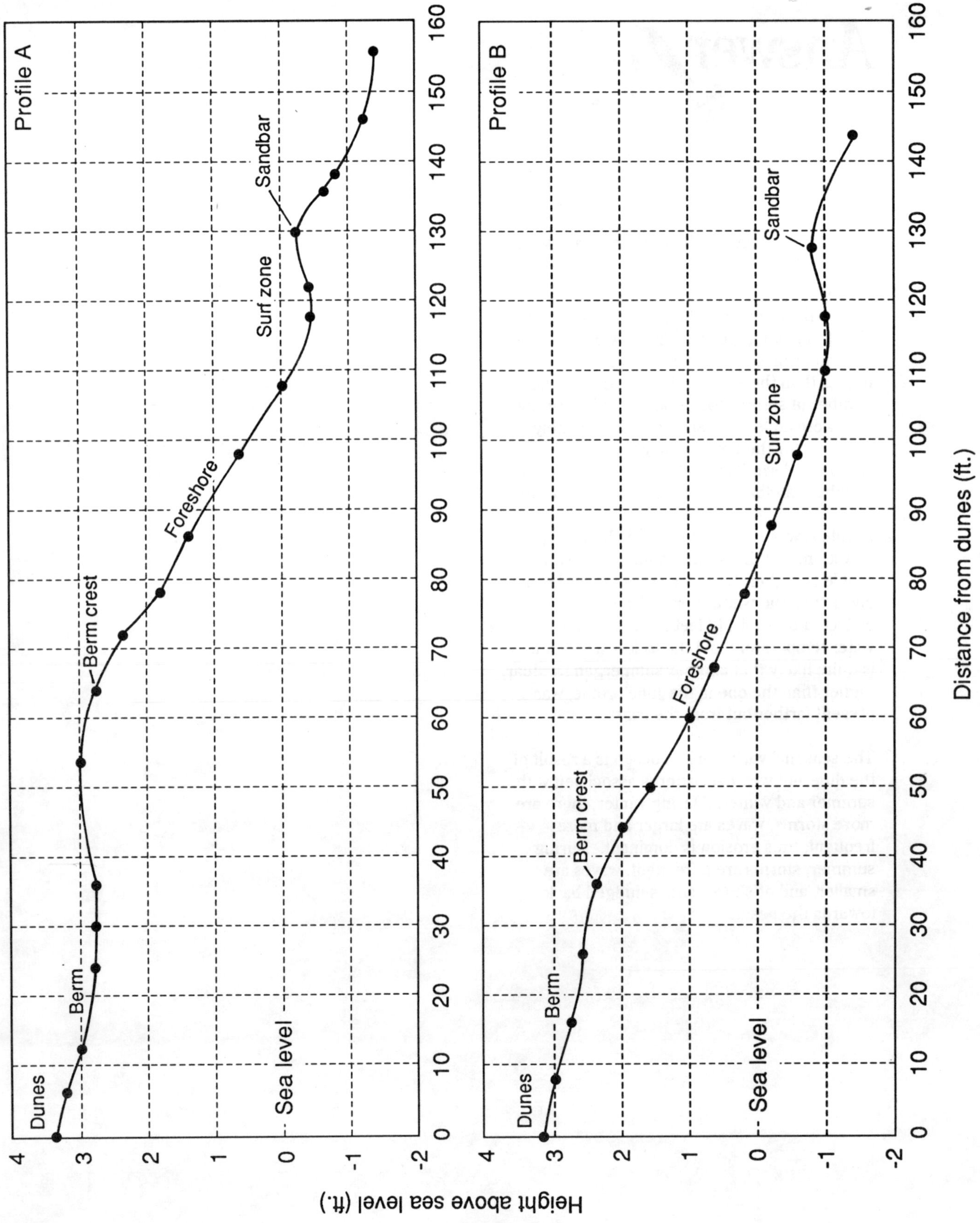

Figure 2. Summer and winter beach profiles of Fire Island, New York.

Building a River

MARYANNE E. PORTER

Academy Park High School
Sharon Hill, Pennsylvania

Objectives

To recognize the stages in the development of a river system from its source to its mouth by constructing a geologic model that simulates river features and processes.

To relate conditions observed along the model river to those described by Mark Twain in *Life on the Mississippi.*

To describe the conditions simulated and observed in the model rivers.

Time required

Activities should be spread over three to four class periods.

Materials

per group of 4-5 students

Stream table: 1.25 m x 30 cm (4 ft. x 1 ft.). Stream tables are available from suppliers such as Ward's Natural Science Establishment, Inc., Rochester, New York.

Water source: two gallons are enough if a circulating pump is used.

Bucket or container for water

Soil: a mix of soil, sand, and pebbles works best. Use enough to fill each stream table about two-thirds full.

Mop (in case of spills)

Life on the Mississippi by Mark Twain: at least one copy

Background

A **river system** consists of a river and all of its tributaries, the streams that flow into it. The following processes and features may occur as a river develops:

bed The bottom of a water course or a stream channel.

channel The bed where a stream or river flows or may flow.

cutoff (a) The new and relatively short channel formed when a stream cuts through the neck of an oxbow or horseshoe bend and thereby shortens its channel. (b) The crescent-shaped body of water separated from the main stream by a cutoff.

delta The nearly flat tract of land at the mouth of a river that results from the accumulation of sediment supplied by the river. Deltas commonly form a triangular or fan-shaped plain.

flood plain The nearly flat portion of a river valley, adjacent to the channel and built of sediments deposited when the river overflows its banks during floods.

headwater The source and upper part of a stream.

meander One of a series of winding curves or loops in the course of a mature stream. A stream produces meanders as it swings from side to side in flowing across its flood plain.

mouth The place of a stream's discharge, as where it enters a larger stream, a lake, or the sea.

neck A narrow strip of land connecting two larger areas.

oxbow A closely looping stream meander that is so extremely curved that only a neck of land separates two parts of the stream.

oxbow lake The crescent-shaped body of standing water that is situated in the abandoned channel of a meander after the ends of the original bend filled with sediment. Examples are common along the banks of the Mississippi River.

sorting of sediments The process by which sedimentary particles of similar size, shape, or weight are naturally separated from dissimilar particles by running water or another agent of transportation.

source The point at which a river begins to flow.

waterfall A perpendicular or steep descent of a stream, as where it crosses an outcrop of resistant rock overhanging softer rock that has been eroded.

Procedure

1. Working in groups, pile soil at one end of a stream table, but spread enough soil to cover about two-thirds of the tray. Incline the stream table 10° or less if features are to form in one class period. If the activity can run longer, a flat tray works best.

2. Start running water from a water source over the soil. The slower the water runs, the better the results will be.

3. Each member of the group should observe and record the formation of any natural occurrences along the river. You will use your observations later to write a description of the model river. Make sketches and diagrams of the feature. For example, a diagram of a meander could show where sediment is deposited and washed away.

4. Use a pencil, or similar object, or your finger to shape the channel so that the following features of the river can be observed: waterfalls, meanders, necks, cutoffs, oxbow lakes, and deltas.

5. Draw the shape of the channel formed and label the features.

6. Read the passages from *Life on the Mississippi* that describe the features of the river, especially the first parts of chapter 17, "Cutoffs and Stephen."

7. Use your drawing and observations to write a description of the river. Compare your description to Mark Twain's description.

8. Follow the guidelines below to revise your description.
   a. Each member of your group will read his or her description to the rest of the group two times.

b. As each person reads for the second time, jot down your thoughts about the description. Put your thoughts in the following categories:

1. What I like about your description is....

2. Questions I still have about your description are....

c. You may use the comments about your description to improve it for a second draft or for the final draft that you will turn in to your teacher.

d. If your group needs help getting started, your teacher may decide to take on the role of a group member.

9. Flood the stream table a few times so that features such as a levee and flood plain form.

Block Diagram Problems

Nancy S. Harms
Pleasant Hill, California

Objectives

To use block diagrams to visualize some simple geologic structures in three dimensions.

To determine the sequence of geologic events leading to the development of the structures and to learn the names of the structures.

Time required

Three class periods to complete 8 to 10 block diagram problems.

Materials

per student
Scissors
Soft-lead pencil
Eraser: cut at an angle to provide a narrow erasing edge (and to double the number of erasers available)
Ruler

Background

Note: Students should have a general knowledge of geologic structures.

Some of our most interesting scenery involves folded and faulted rock layers. It is not difficult to learn what various geologic structures look like from all sides. This investigation will guide you through a series of diagrams that become more and more complex. There are simple rules you must follow, and you will work with drawing techniques you might not have encountered elsewhere. When you follow these rules through every step of your work, you are assured of a correct solution. As you work on this activity, you will develop your eye-hand coordination and polish your logical, sequential thinking skills.

Procedure

In this investigation, block diagrams represent geologic structures. You will be given the pattern of one or more faces of a structure. Determine what the other faces will look like. Use information in Parts I to IV in the **Procedure** and the symbols shown in Figure 1 to help you figure out and draw the pattern on the unmarked faces.

Use a soft-lead pencil for all drawing (NO pens!); pencils must be kept sharp. Number all diagrams. Do the assigned diagrams in numerical order. Do not skip around.

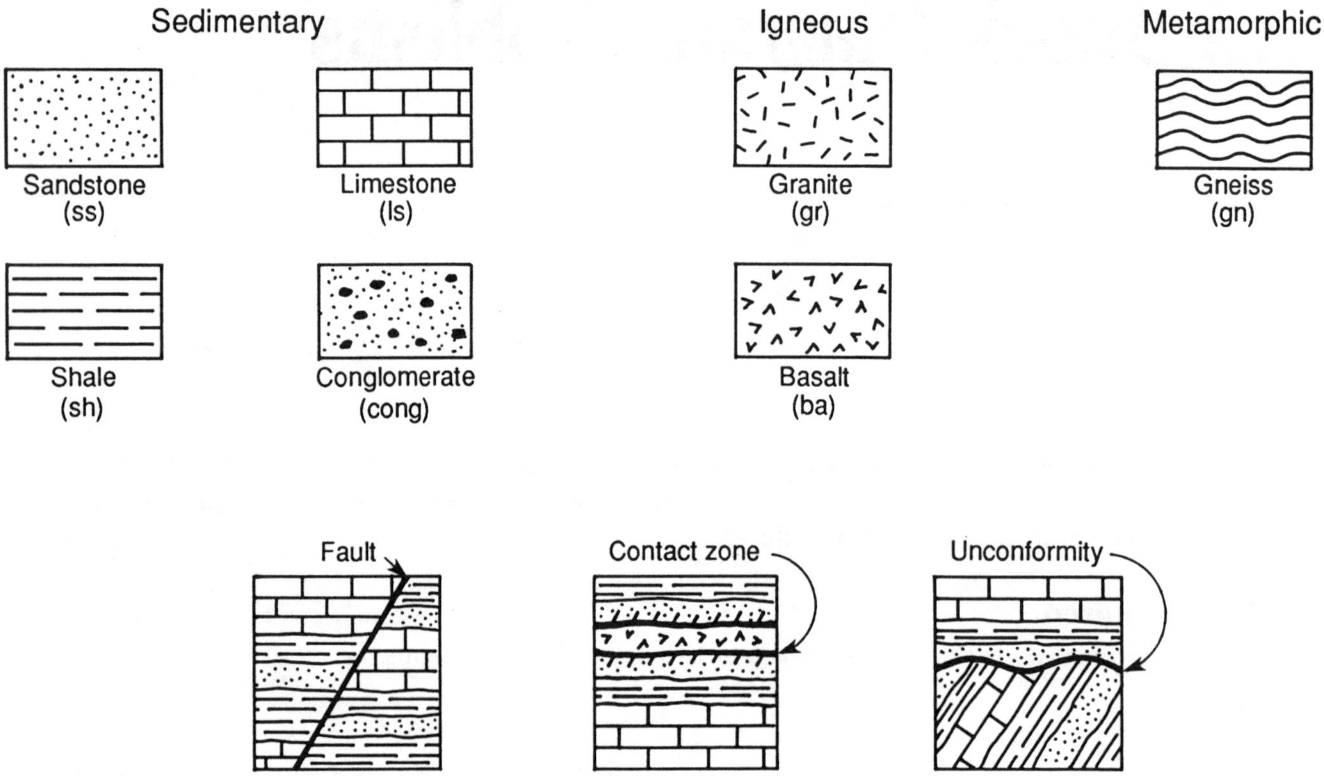

Figure 1. Symbols for rock types and structures.

Part I. Assumptions

You will need to make the following assumptions to figure out how to complete each block diagram.

1. The pattern that appears along the edge of one face will also appear along the edge of the adjoining face unless otherwise indicated.

2. The pattern on one face will appear on the opposite face unless otherwise indicated.

3. Faults will extend through the block.

4. Folds will extend through the block.

5. Rock layers will extend through the block unless otherwise indicated.

Part II. Folding and drawing

Follow the steps below for each block diagram. See Figure 2.

1. Cut the diagram out. Make it into a box by folding each of the four flaps along the outside edge of the center square so that the patterned flap faces out (Figure 2A-C). Make very sharp creases. Do not tape or staple the flaps because you will need to have the paper flat later. Correct folding is essential.

2. Hold the box from the inside so that one blank flap (face) is next to the patterned face (Figure 2C).

3. On two blank faces that adjoin the patterned face (you are holding one blank face, and the other is the top face), draw short horizontal lines to mark the top and bottom of each of the layers shown on the patterned face (Figure 2D).

4. Use the assumptions in Part I to help you decide the direction the rock layers go, and draw lines to show the direction on the two faces (Figure 2E).

5. Draw the appropriate rock pattern for each layer (Figure 2F).

6. Fold, mark, and draw until all the faces have been patterned. The solution to each diagram is easier when you hold the box in one hand and view it from all angles. Check your work, making sure layers continue around corners and that no layer ends abruptly.

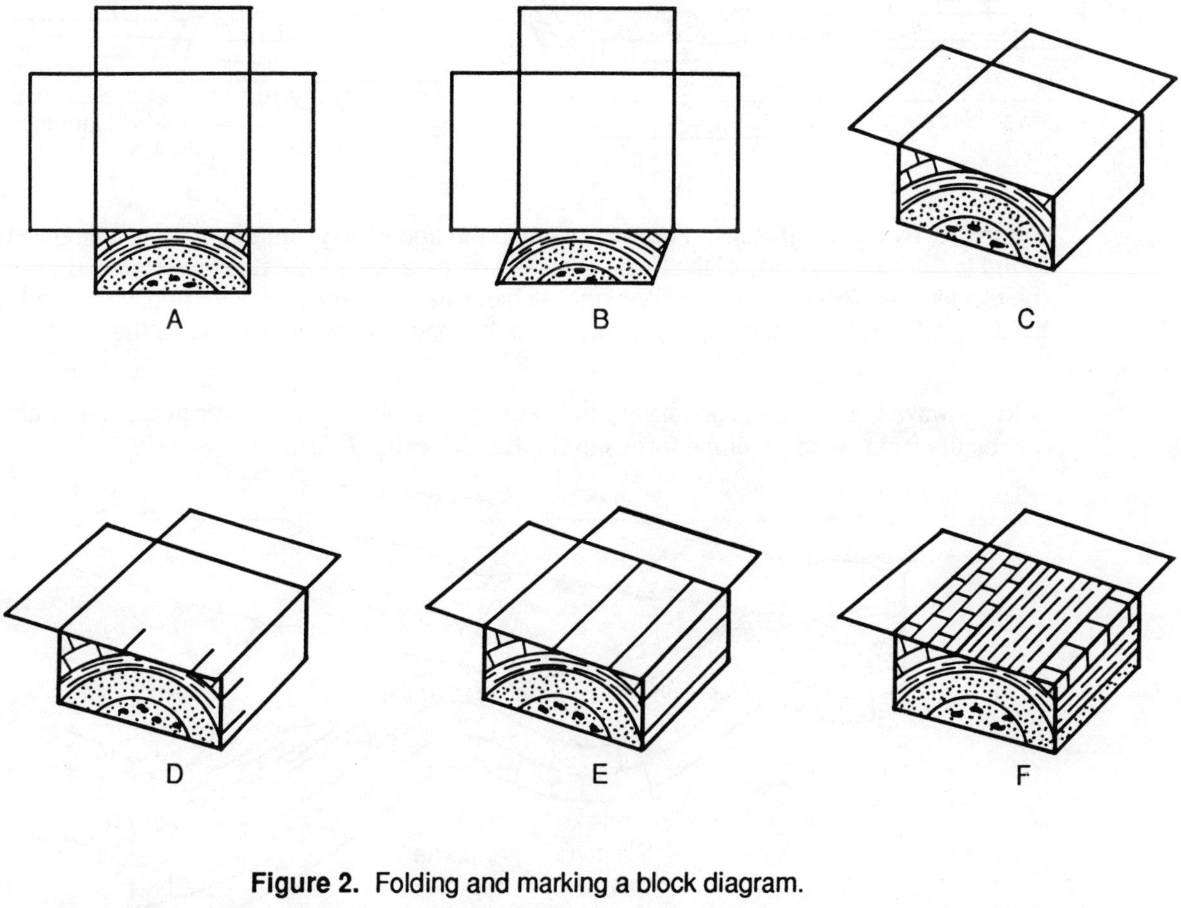

Figure 2. Folding and marking a block diagram.

Part III. Vocabulary

You will need to understand the following vocabulary.

anticline Upturned fold. See "fold" (Figure 4).

contact metamorphism The alteration of rock by heating due to contact or proximity to molten rock. New minerals are commonly formed, and the texture of the rock is changed.

contact zone The area around an igneous rock, which shows the effects of contact metamorphism.

deposition The natural process of laying down rock-forming material.

discontinuity A break or time gap in the deposition of layers.

erosion The natural process of moving rock material by water, wind, ice, or gravity.

fault A fracture along which the rocks on one side of the break have moved with respect to the rocks on the other side of the break. Two types of faults are shown in Figure 3.

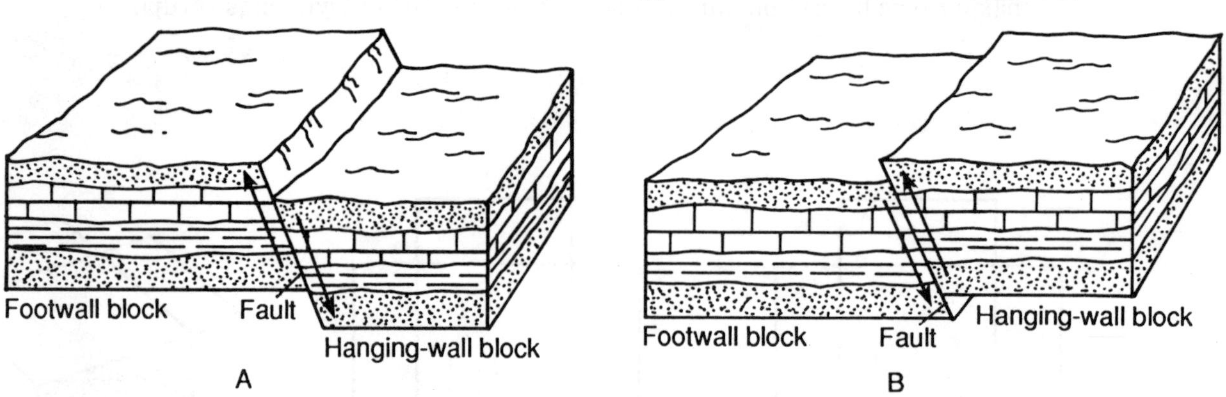

Figure 3. Two types of faults. Footwall refers to the underlying side of the fault. Hanging wall refers to the overlying side of the fault. Direction of the arrows indicates the relative movement of the blocks. (A) Normal fault, the hanging-wall block has moved downward relative to the footwall block. (B) Reverse fault, the hanging-wall block has moved upward relative to the footwall block.

fold A wave-like bend in rock layers that were originally horizontal, or nearly so. Folding results from compressional forces in the Earth's crust (Figure 4).

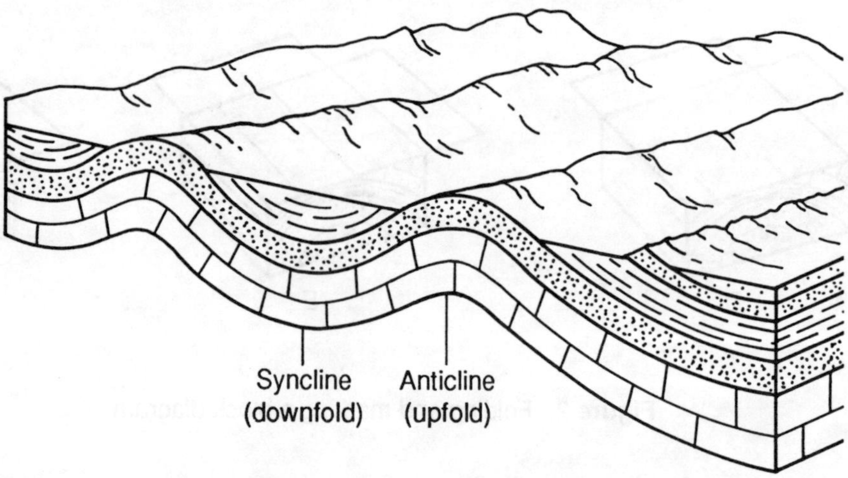

Figure 4. Two types of folds. Syncline, youngest rocks are in the core; anticline, oldest rocks are in the core.

horizontal Parallel to the horizon.

igneous rocks Rocks that are formed from the solidification of magma.

intrusion Magma that pushes into, through, or between existing rock layers or sediments beneath the Earth's surface and solidifies to form igneous rock (Figure 5).

Figure 5. This type of intrusion is called a sill.

lava flow Magma that flows onto the surface of the Earth as a result of volcanic activity and solidifies.

magma Molten or partly molten rock material.

metamorphic rocks Rocks that are changed by heat, pressure, the presence of chemically active fluids, or a combination of the three, without melting.

sedimentary rocks Layered rocks that are formed by the accumulation and consolidation of rock particles or debris. Some sedimentary rocks are formed from chemical precipitates.

sill A tabular igneous intrusion that is parallel to the rock layers surrounding it (Figure 5).

subsidence The sinking or downward settling of the Earth's surface.

syncline Downturned fold. See "fold" (Figure 4).

tabular A shape that is similar to that of a table in which the thickness is small relative to the length and width.

tilted Displaced from the horizontal; at an angle.

unconformity A buried surface of erosion or nondeposition that separates younger from older rock. An unconformity usually results when rock layers have undergone a sequence of uplift, erosion, subsidence, then deposition.

uplift A structurally high area in the Earth's crust, produced by movements that raised the rocks, as in a broad dome or arch.

vertical Upright; straight up and down.

Part IV. Transferring angles

Follow the steps below to transfer an angle from one face to the opposite face. Figure 6 shows how to transfer a fault angle.

1. Place one edge of a ruler at the point where the fault line on the patterned face meets the top edge of the block. Make sure the ruler is parallel to the left edge of the top of the block. Mark the top edge of the opposite face with a light pencil mark (Figure 6B).

2. Move the ruler to the point where the fault line meets the lower edge of the face. Again, keep the ruler parallel to the left edge of the top of the block. Mark the point on the bottom edge of the opposite face (Figure 6C).

3. Draw a line to connect the two points (Figure 6D). You have constructed a fault line that has the correct angle but is of opposite inclination from the angle on the opposite face. When the diagram is folded, the fault line will look as if it goes through the block. Other points can be transferred across the diagram using this method.

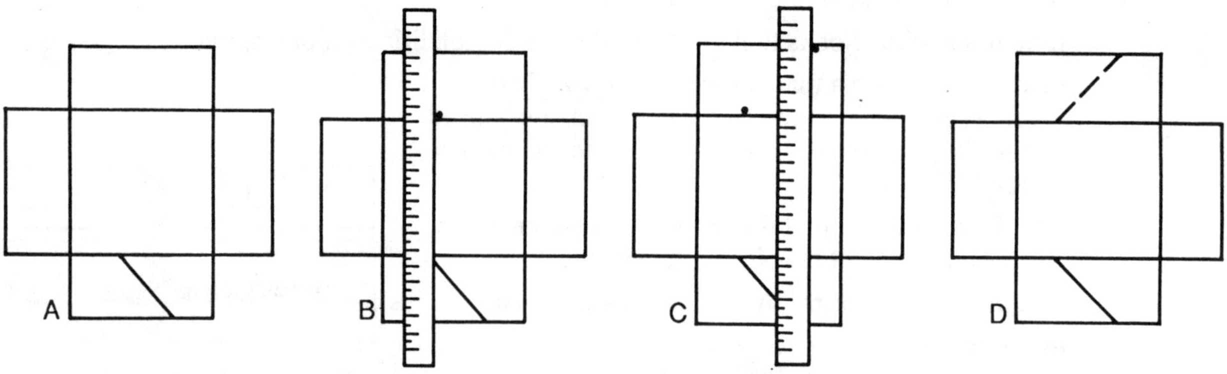

Figure 6. Transferring a fault angle to the opposite face.

Additional resources

Judson, Sheldon, and Kauffman, Marvin E. *Physical Geology*. 8th ed. Englewood Cliffs, N.J.: Prentice Hall, 1990.

Press, Frank, and Siever, Raymond. *Earth*. 4th ed. New York: W.H. Freeman and Co., 1986.

Name ....
Class ....
Date ....

Worksheet 1

Diagram 1

Worksheet 2

Name ....

Class ....

Date ....

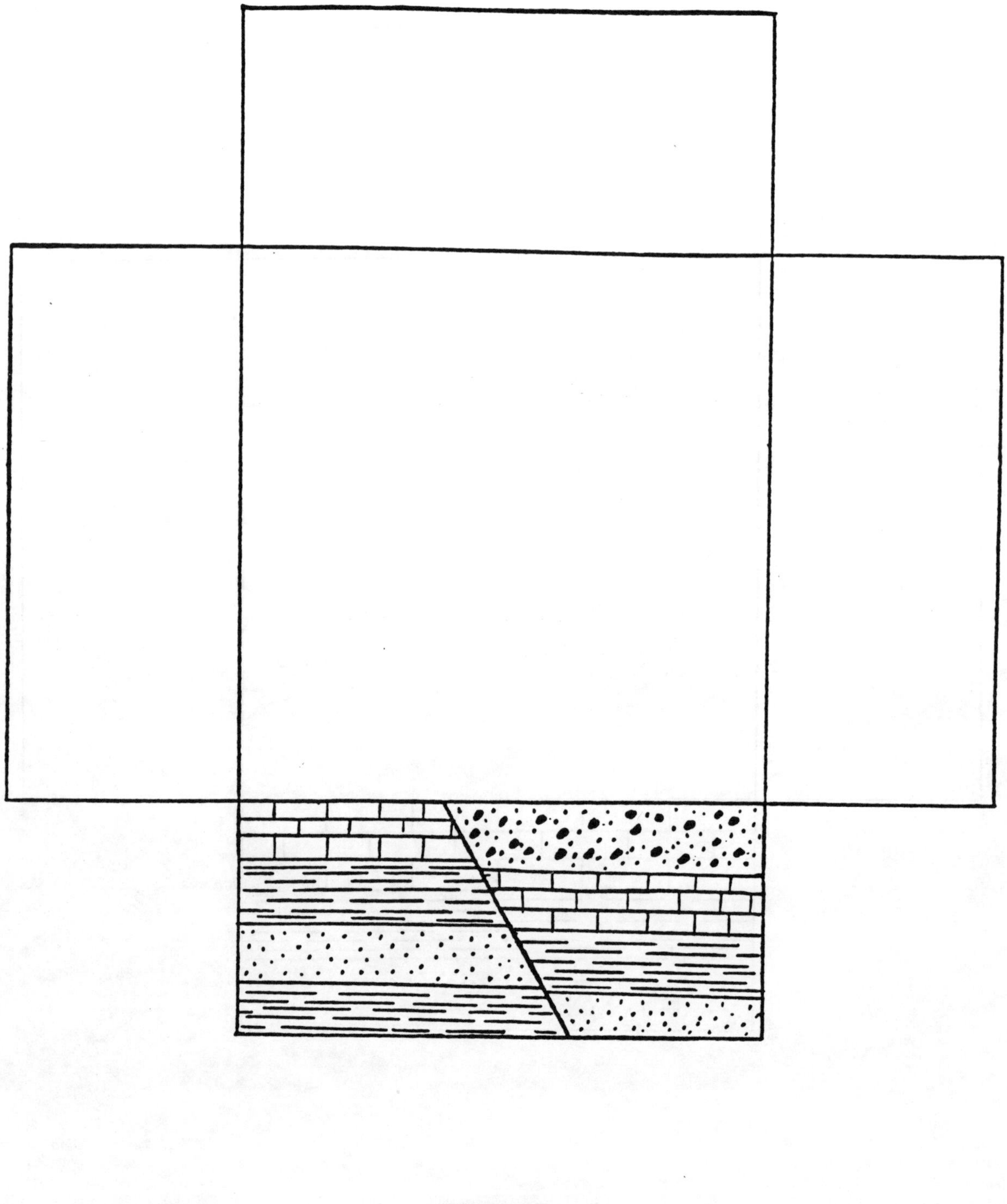

Diagram 2

Name ....

Class ....

Date ....

Worksheet 3

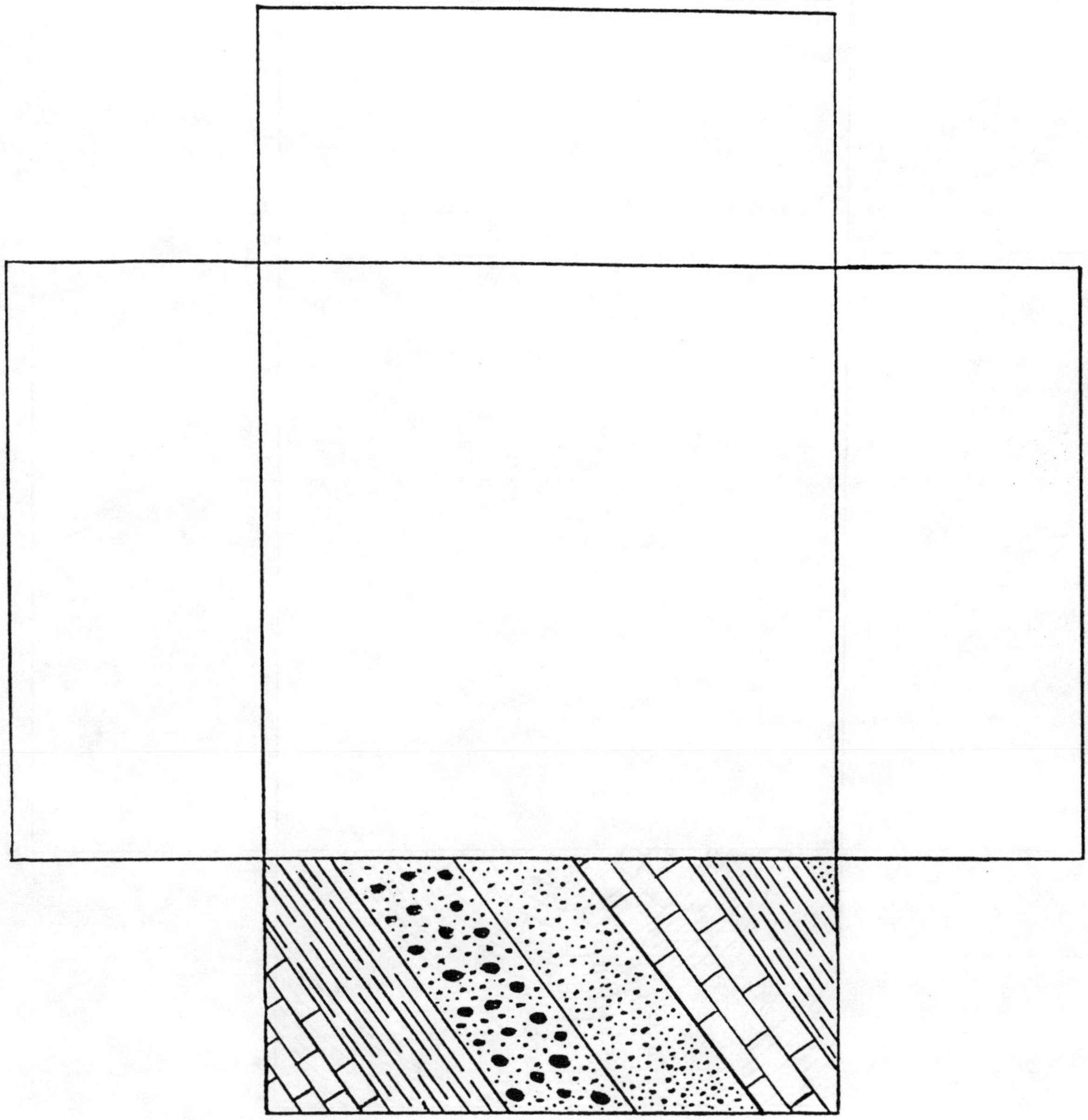

Diagram 3

Worksheet 4

Name ....
Class ....
Date ....

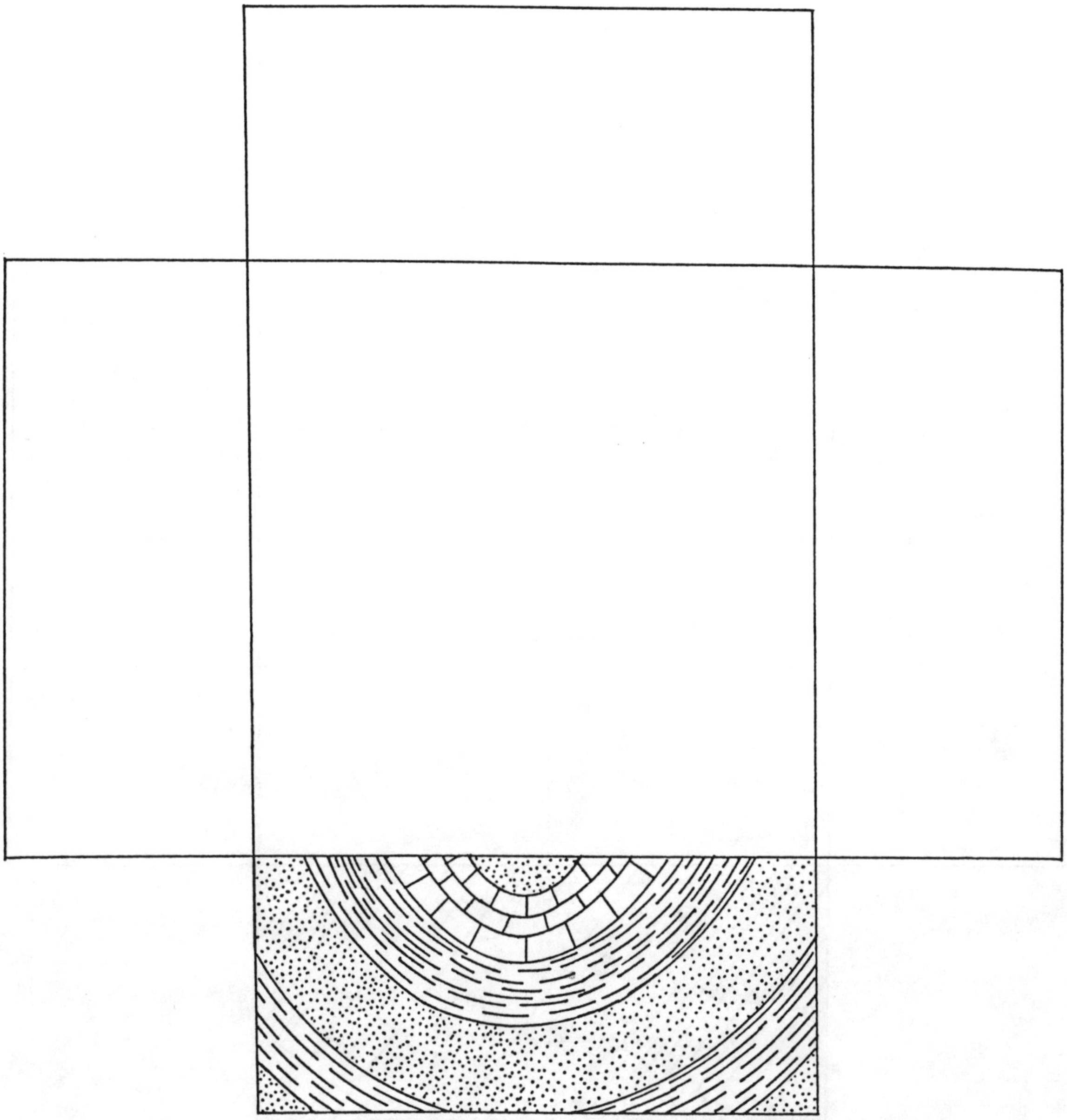

Diagram 4

Name ....
Class ....
Date ....

Worksheet 5

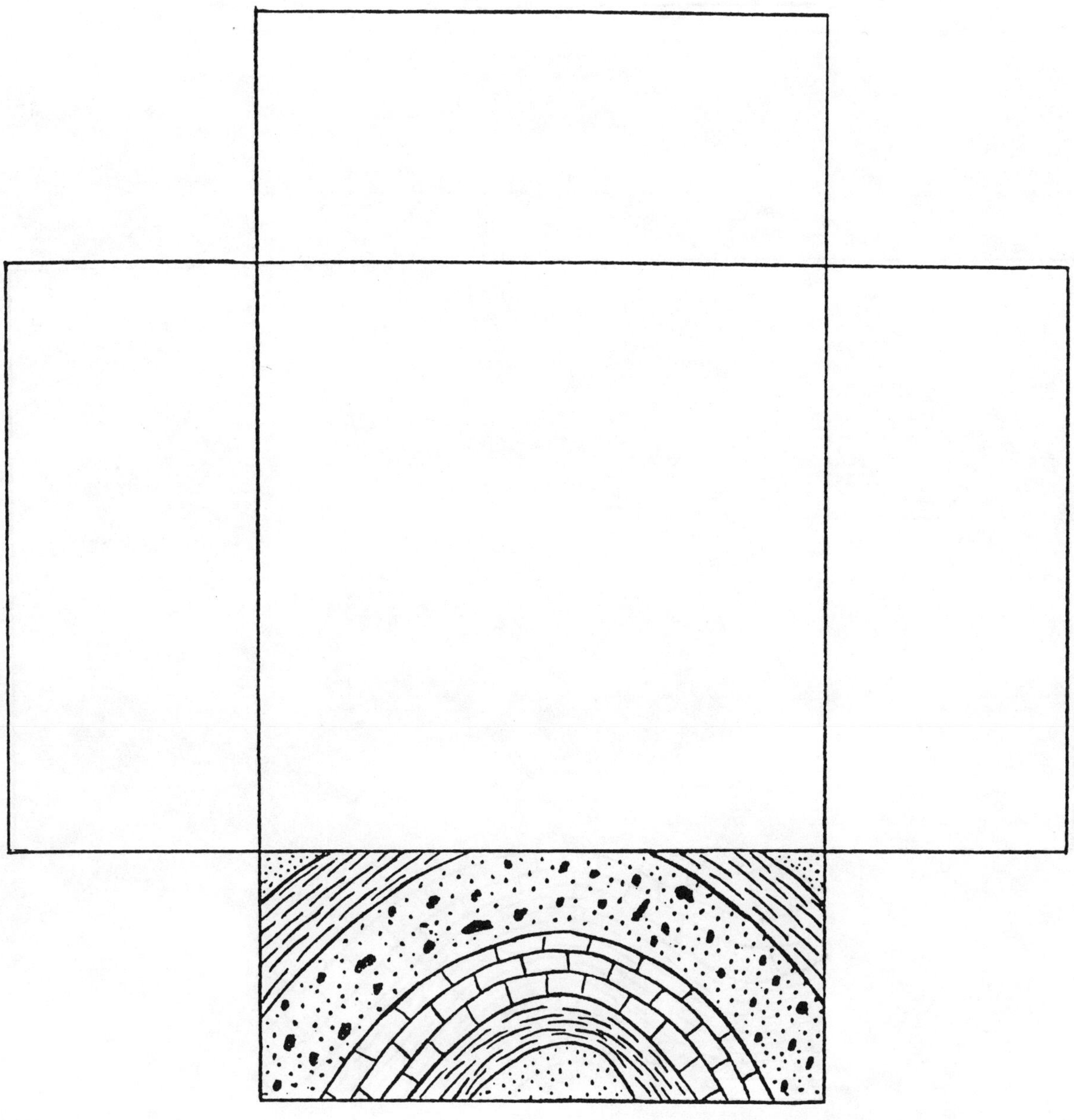

Diagram 5

Worksheet 6

Name ....

Class ....

Date ....

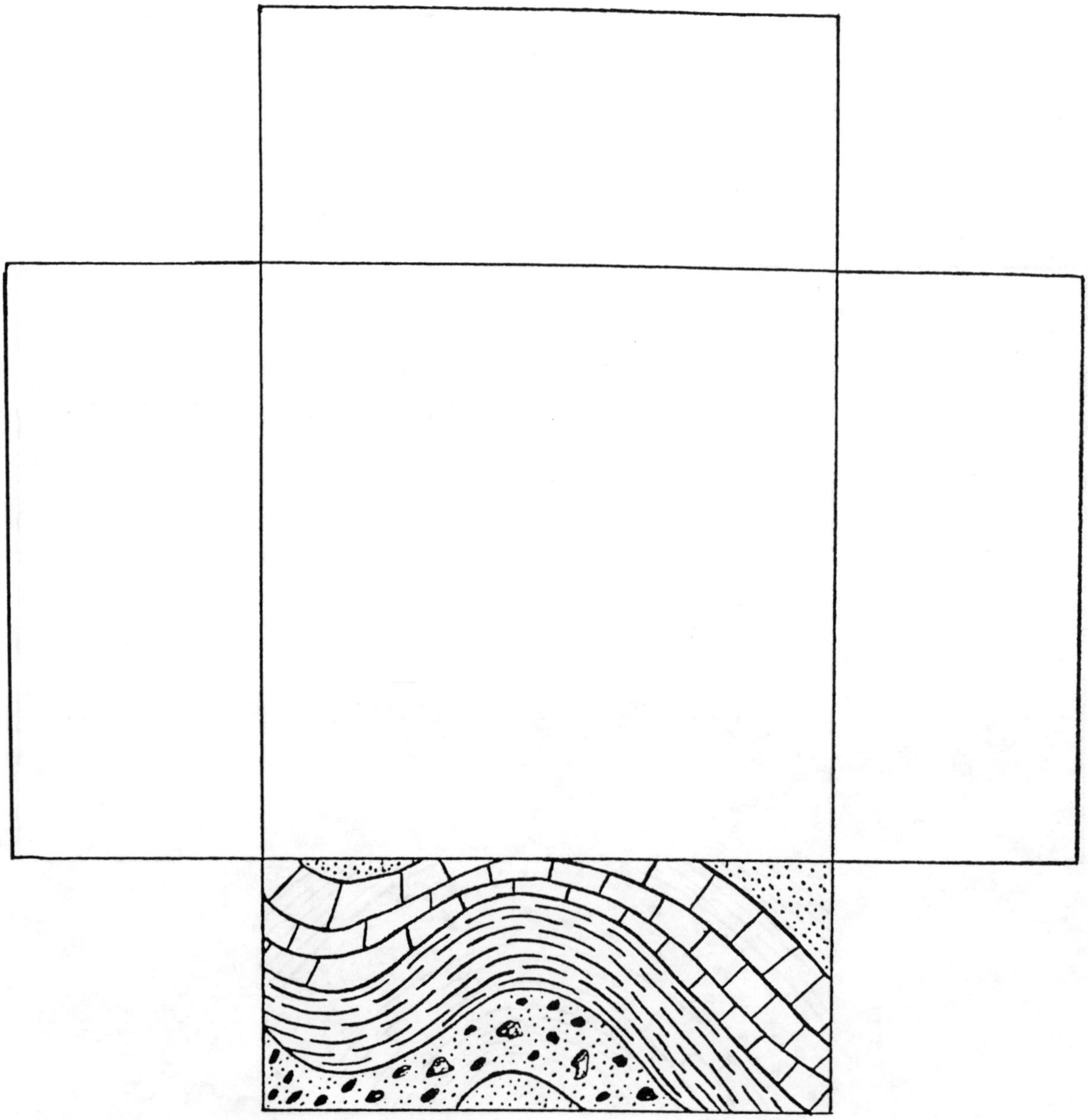

Diagram 6

Name ....

Class ....

Date ....

Worksheet 7

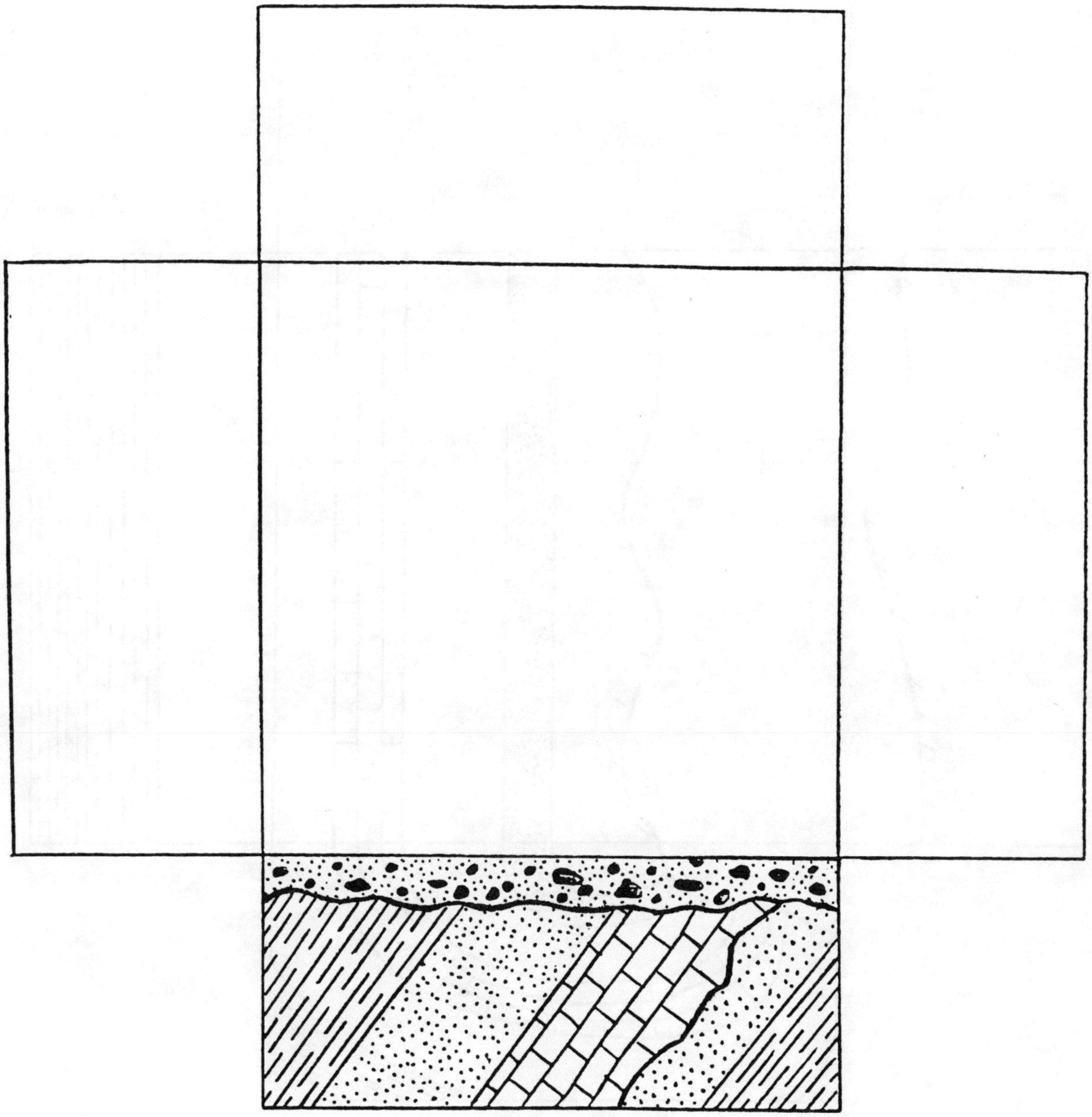

Diagram 7

Worksheet 8

Name ....

Class ....

Date ....

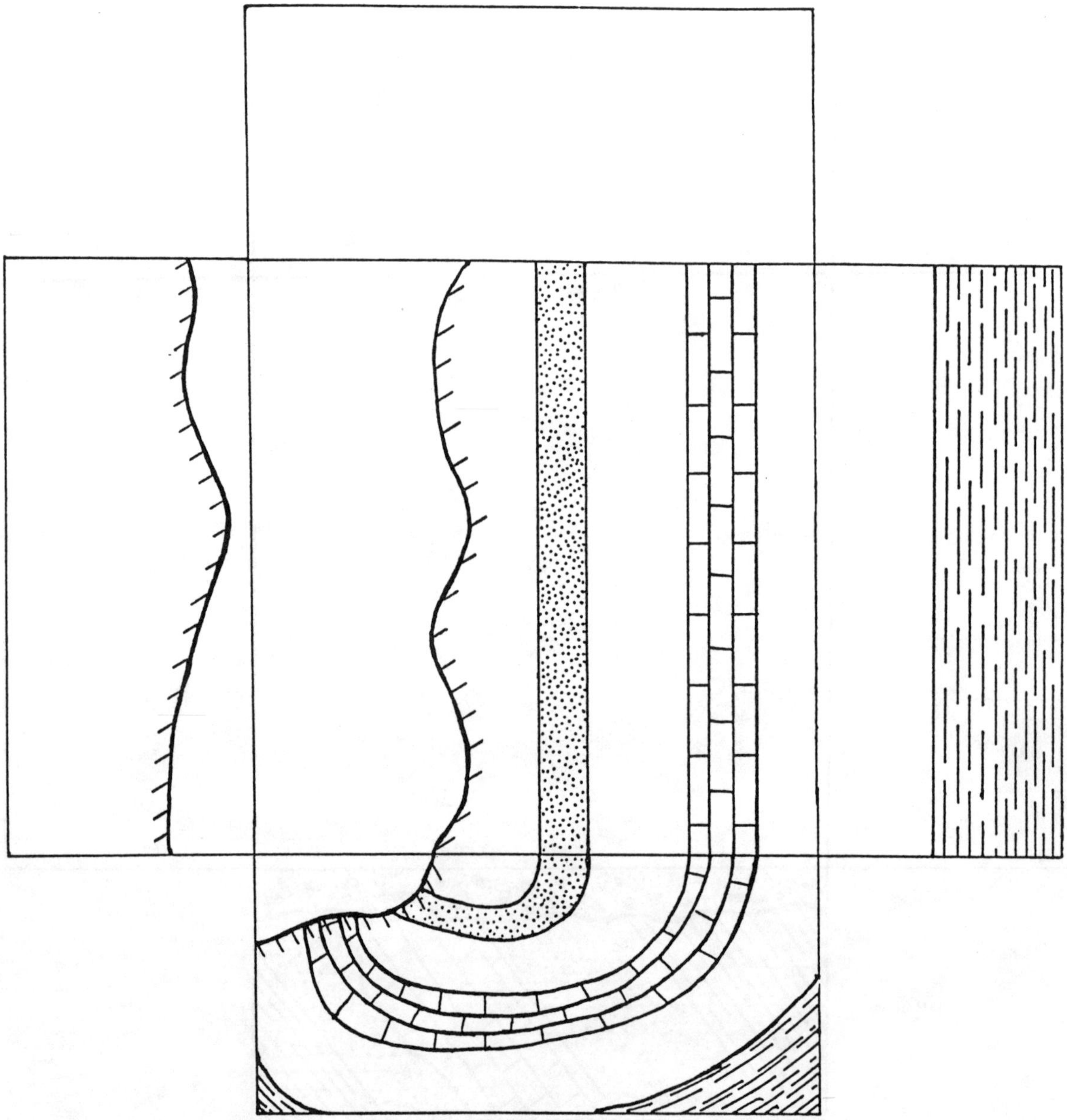

Diagram 8

Name ....

Class ....

Date ....

Worksheet 9

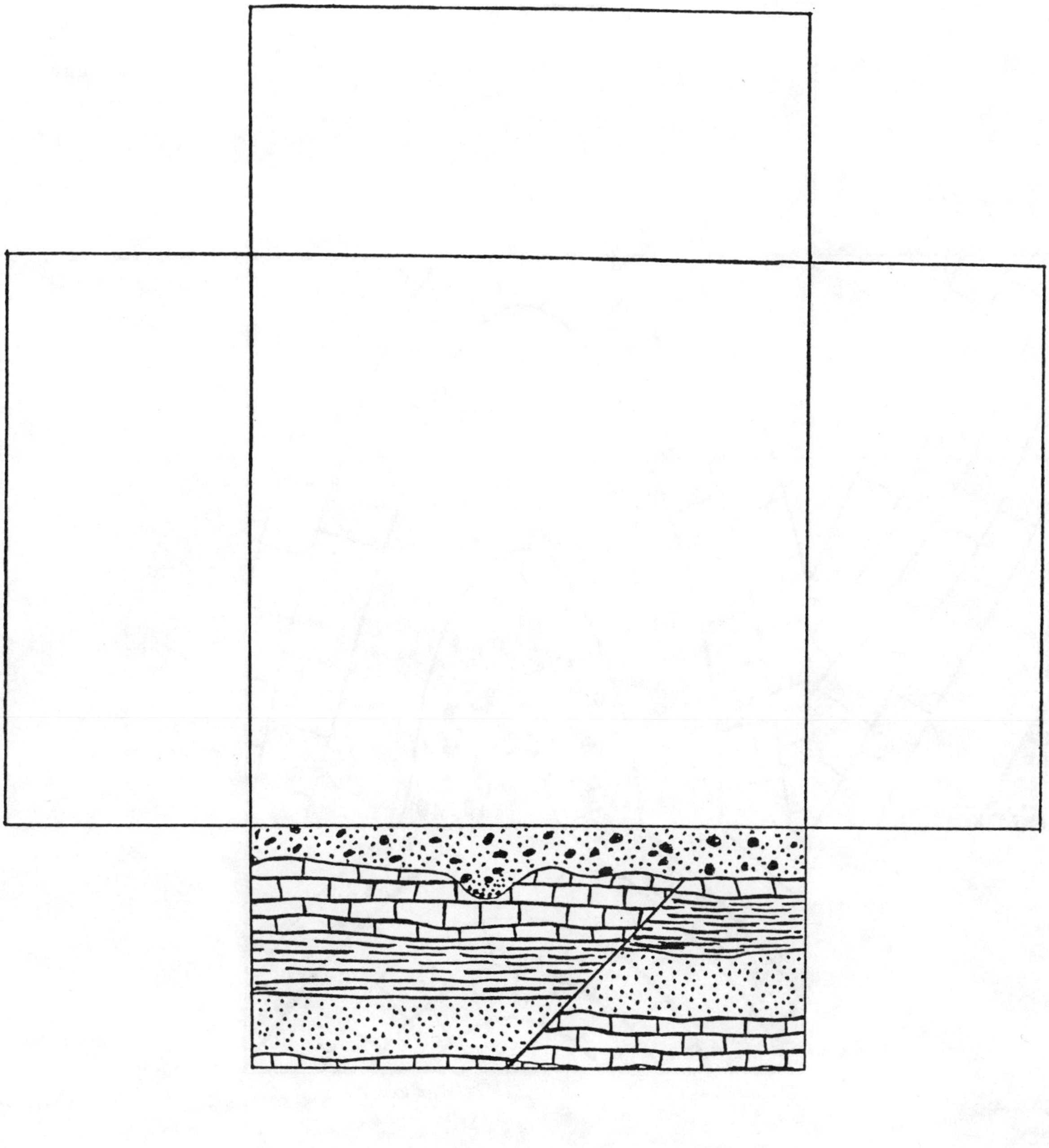

Diagram 9

Worksheet 10

Name ....

Class ....

Date ....

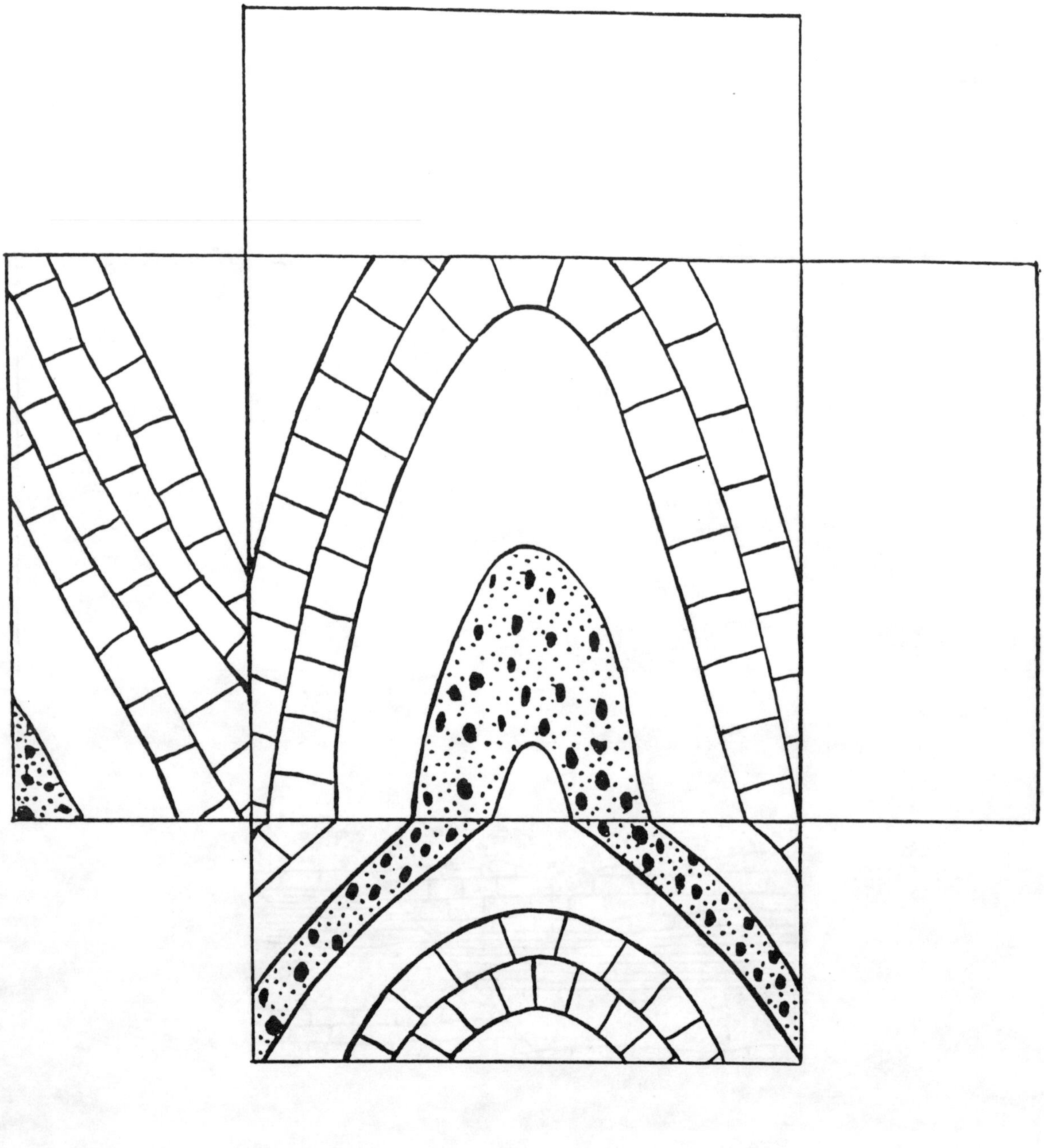

Diagram 10

Name ....

Class ....

Date ....

Worksheet 11

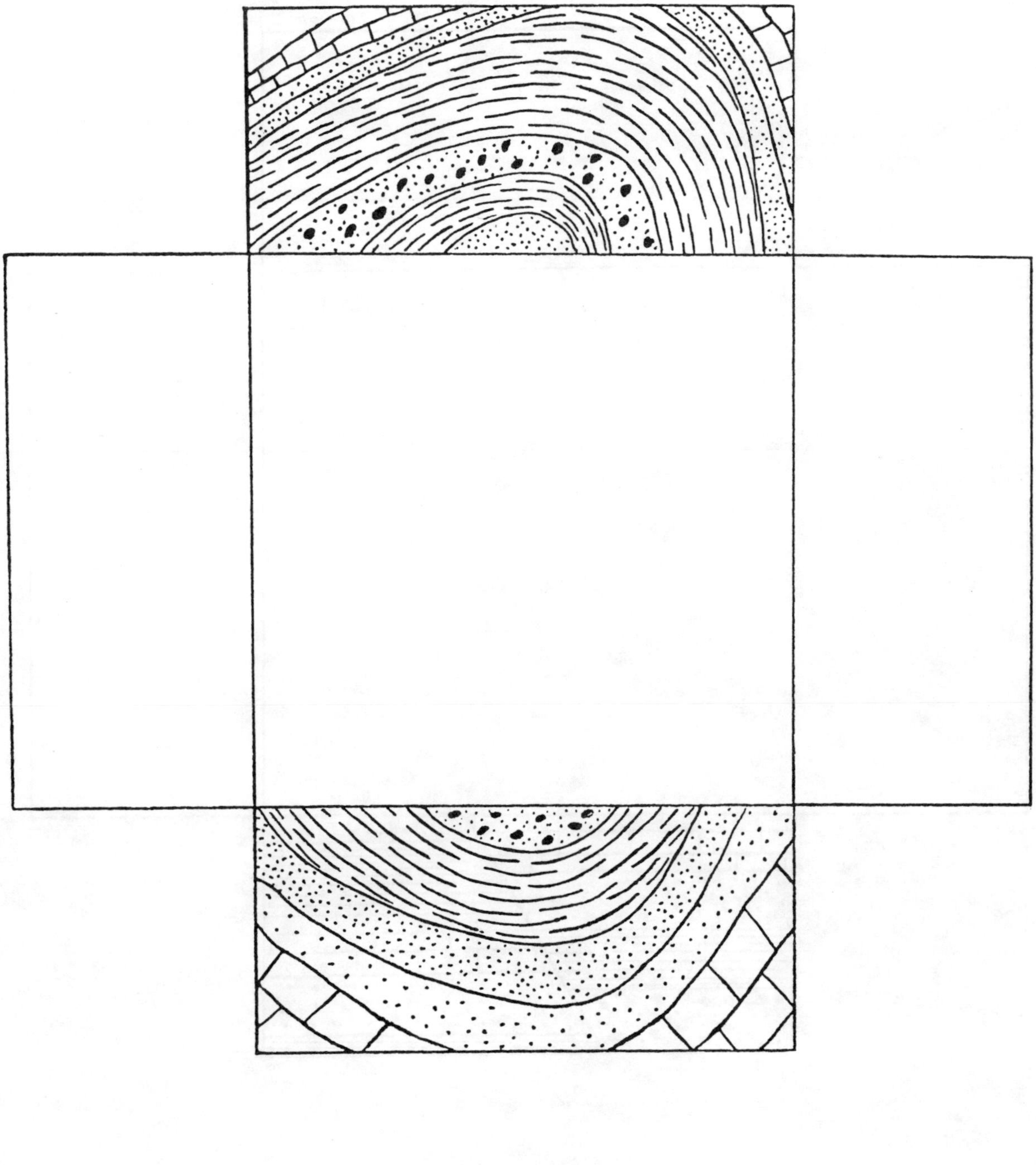

Diagram 11

Worksheet 12

Name ....

Class ....

Date ....

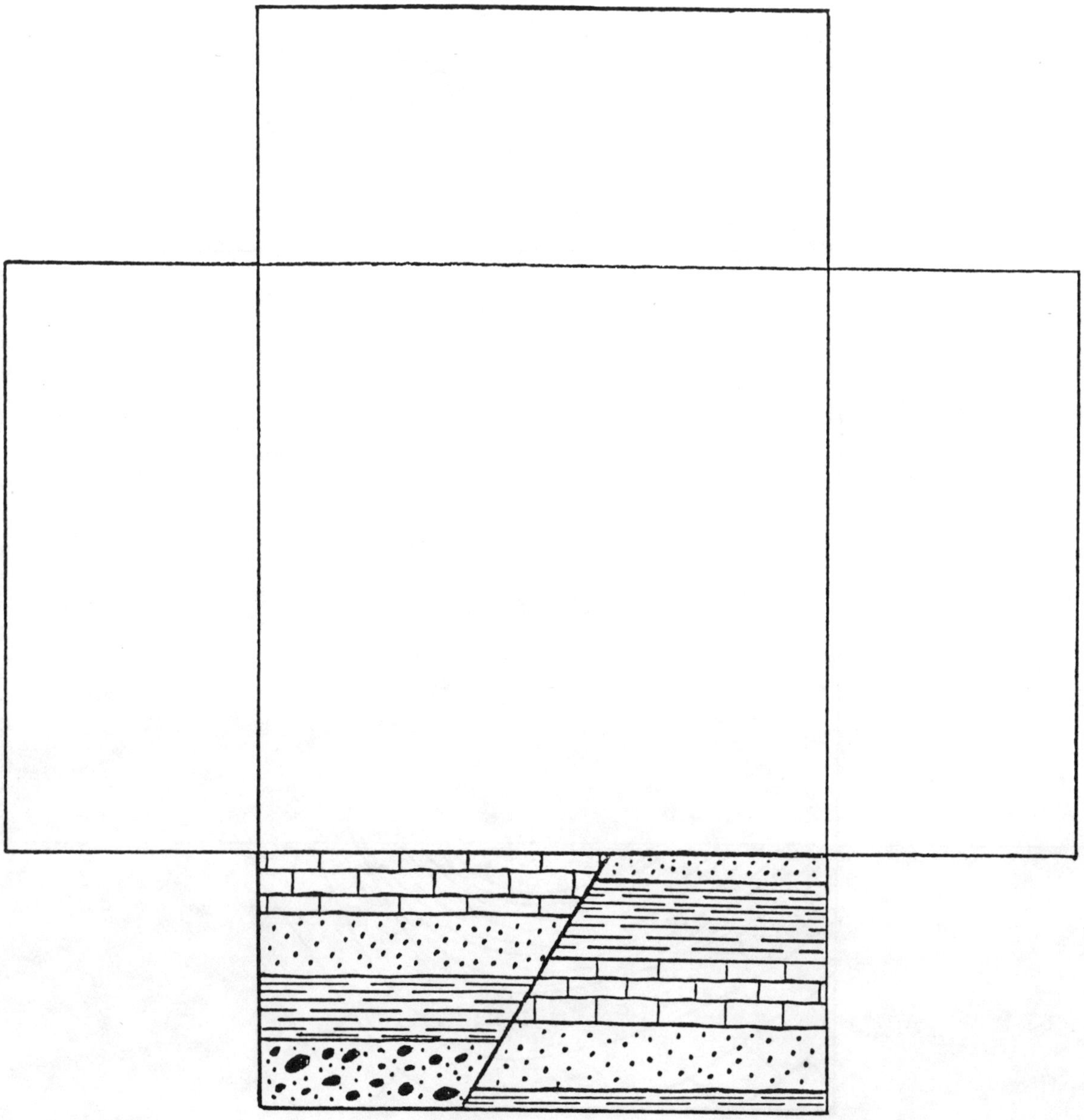

Diagram 12

Name ....

Class ....

Date ....

Worksheet 13

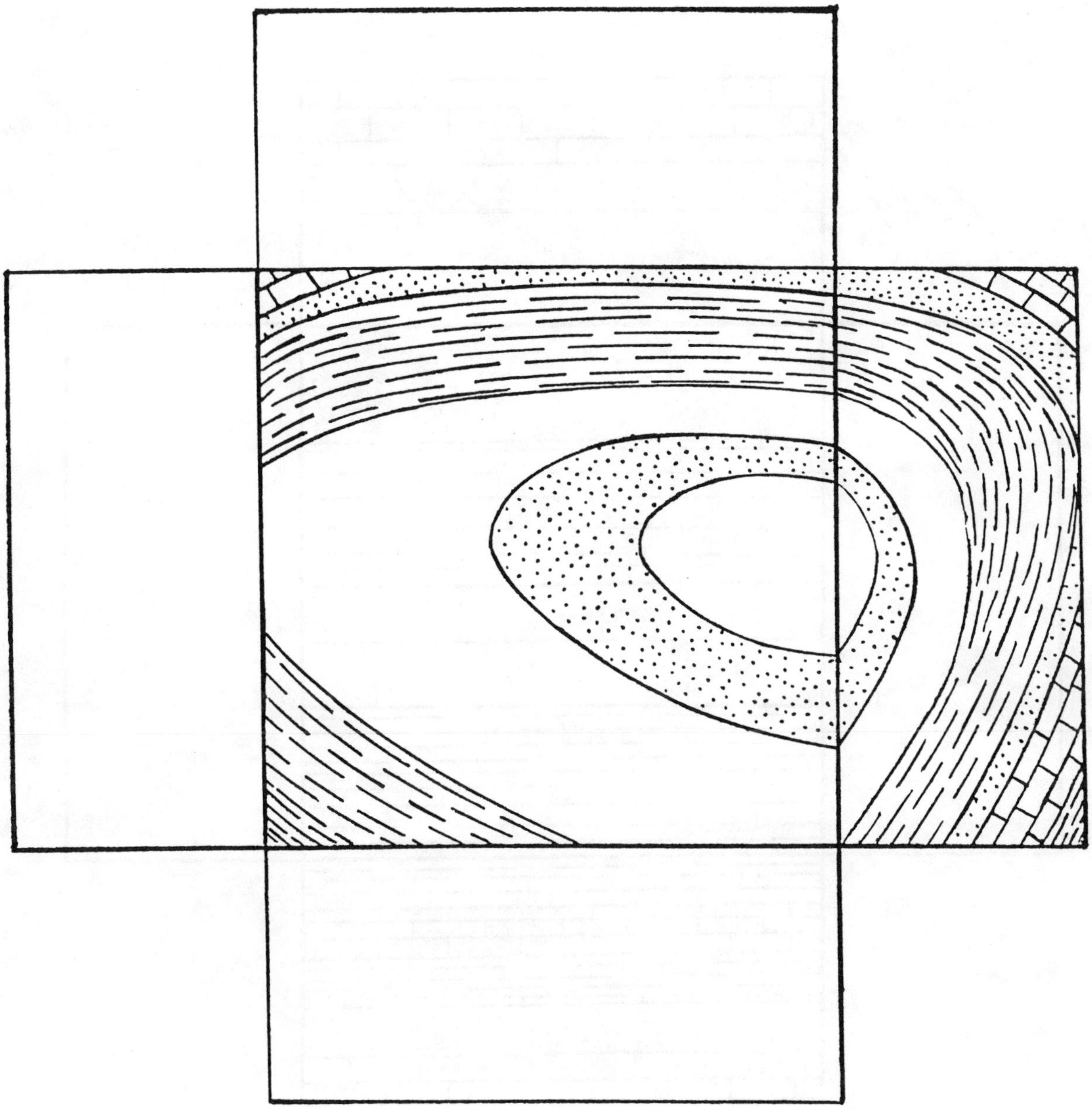

Diagram 13

Worksheet 14

Name ....

Class ....

Date ....

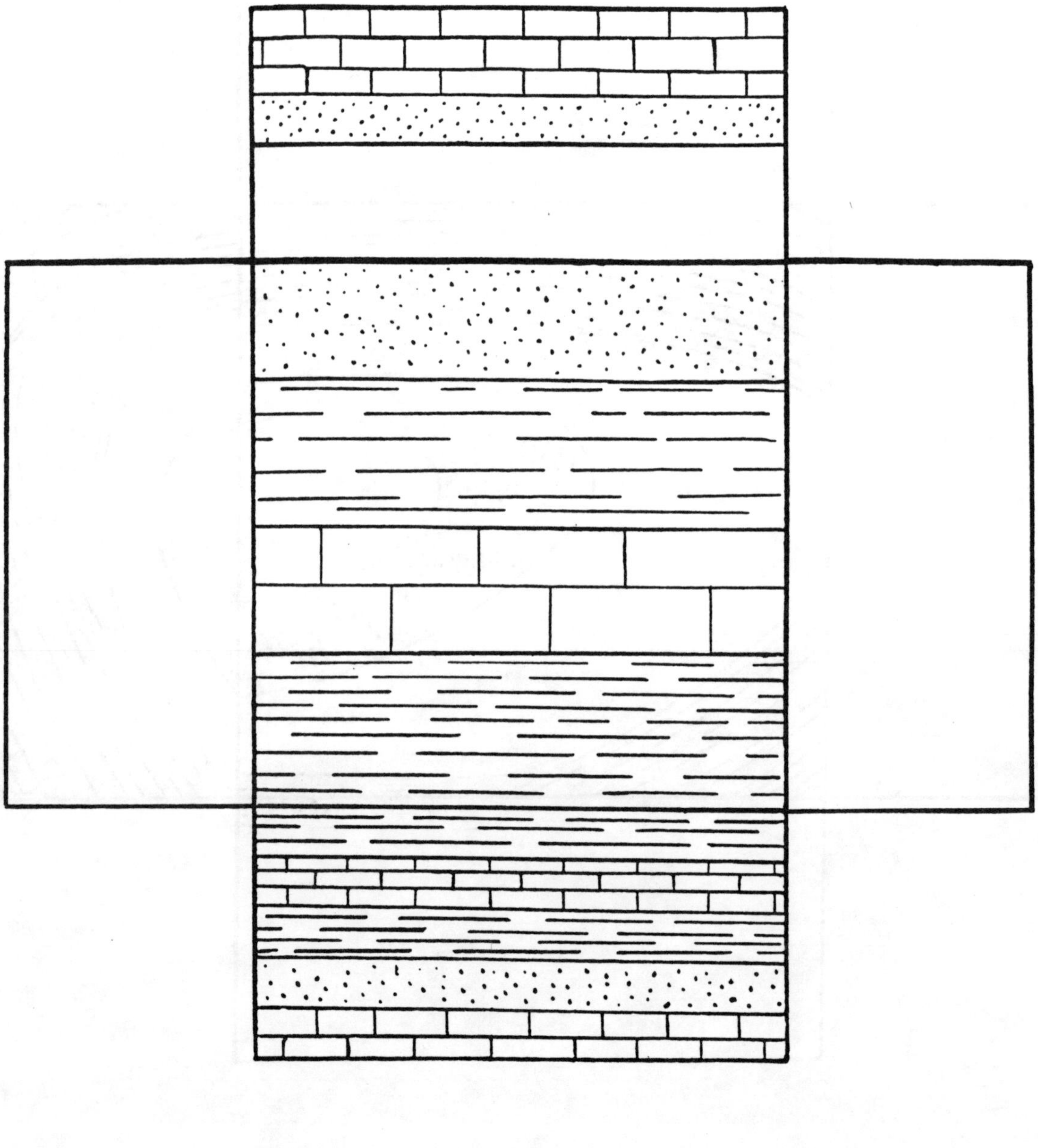

Diagram 14

Name ....

Class ....

Date ....

Worksheet 15

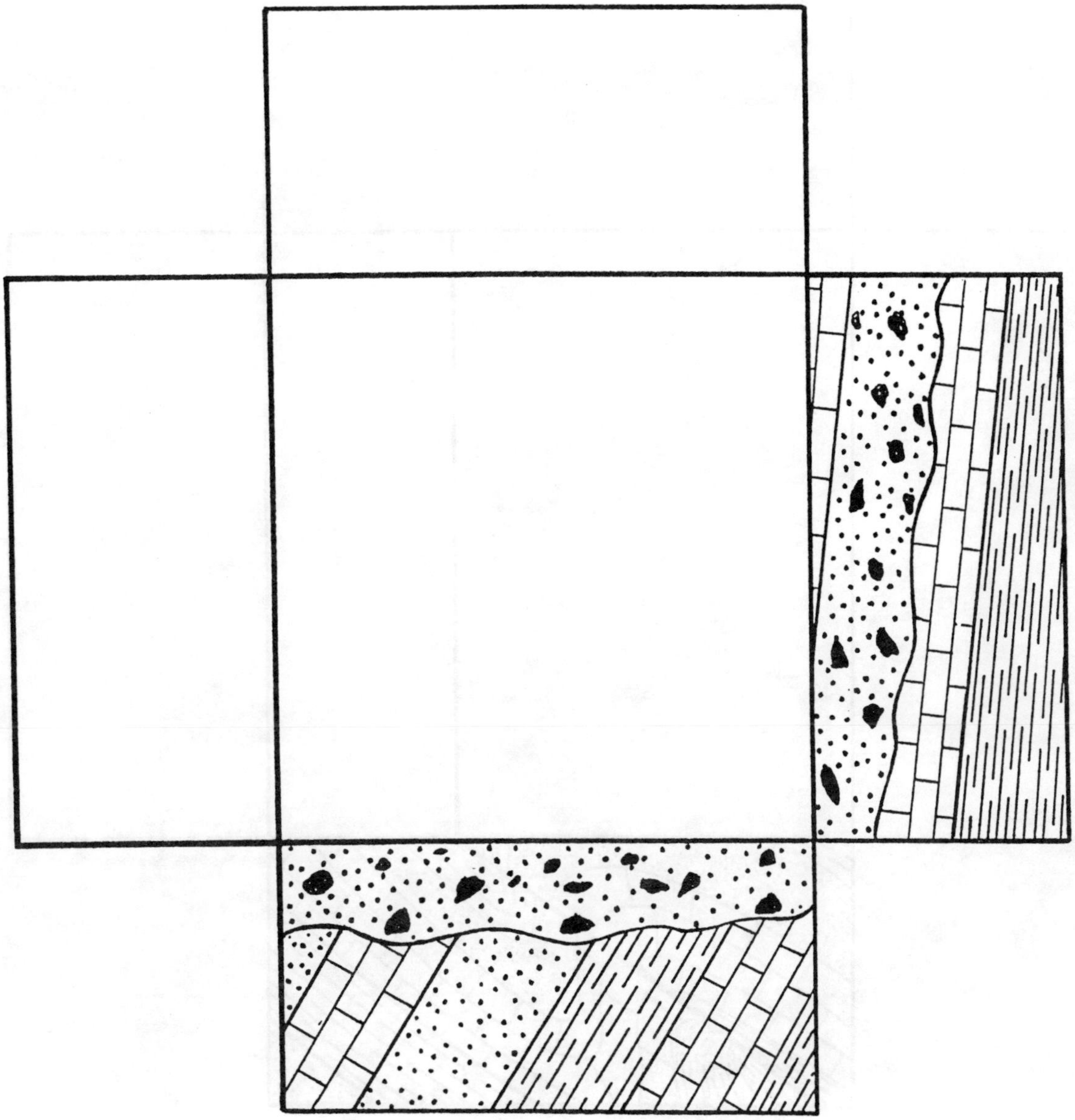

Diagram 15

Worksheet 16

Name ....

Class ....

Date ....

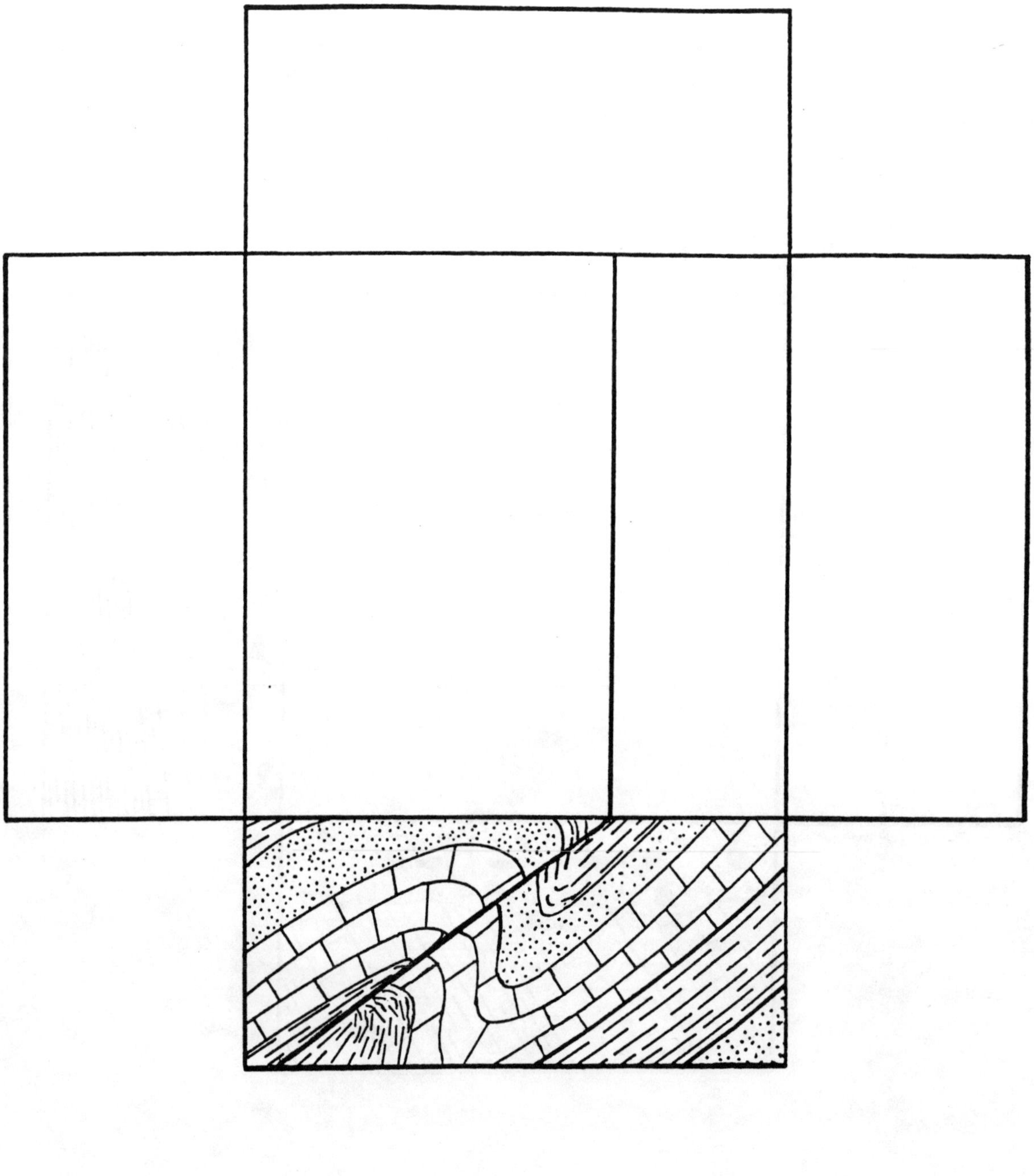

Diagram 16

Name ....

Class ....

Date ....

Block Diagram Problems

Worksheet 17

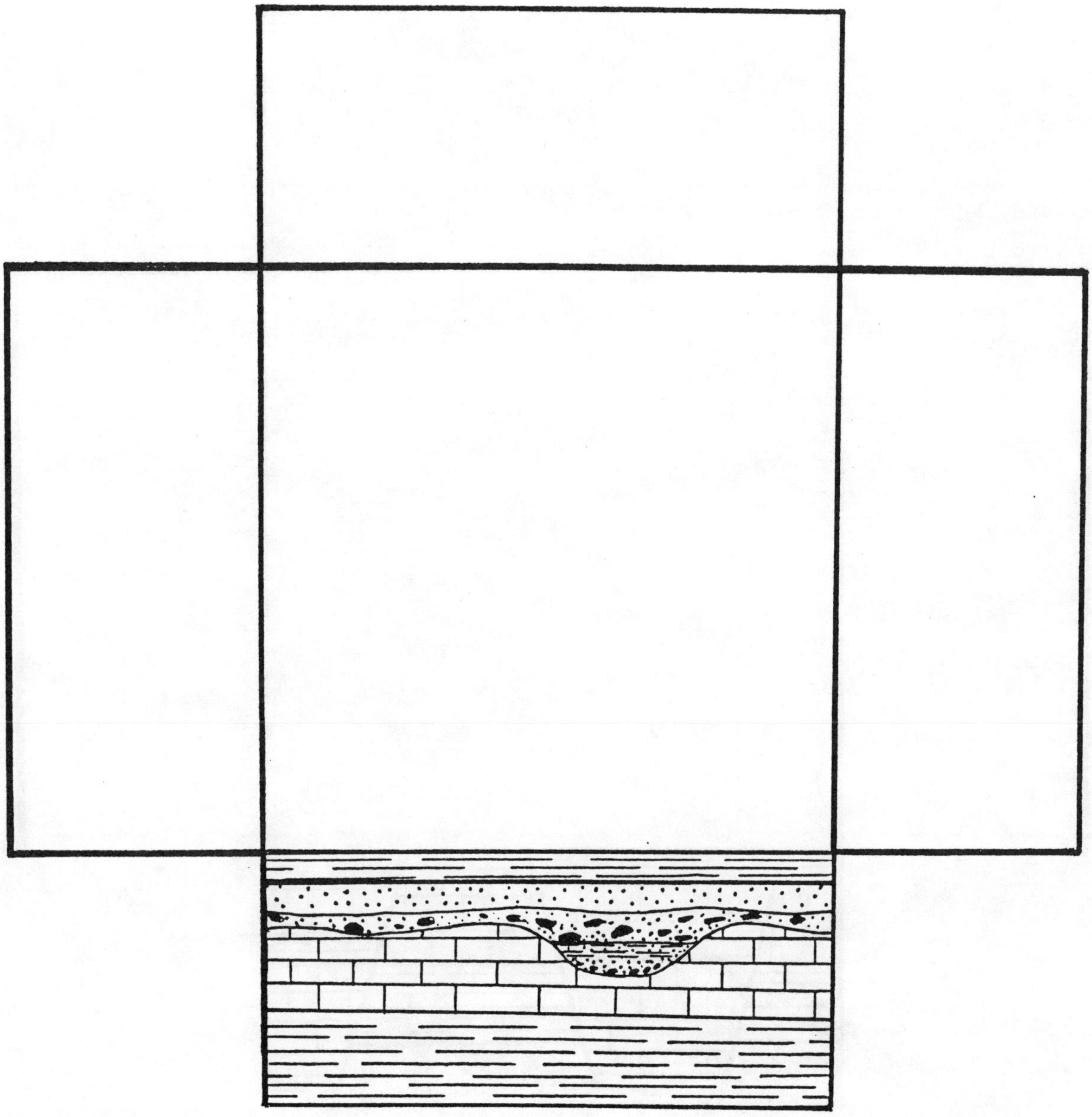

Diagram 17

Block Diagram Problems

Worksheet 18

Name ....

Class ....

Date ....

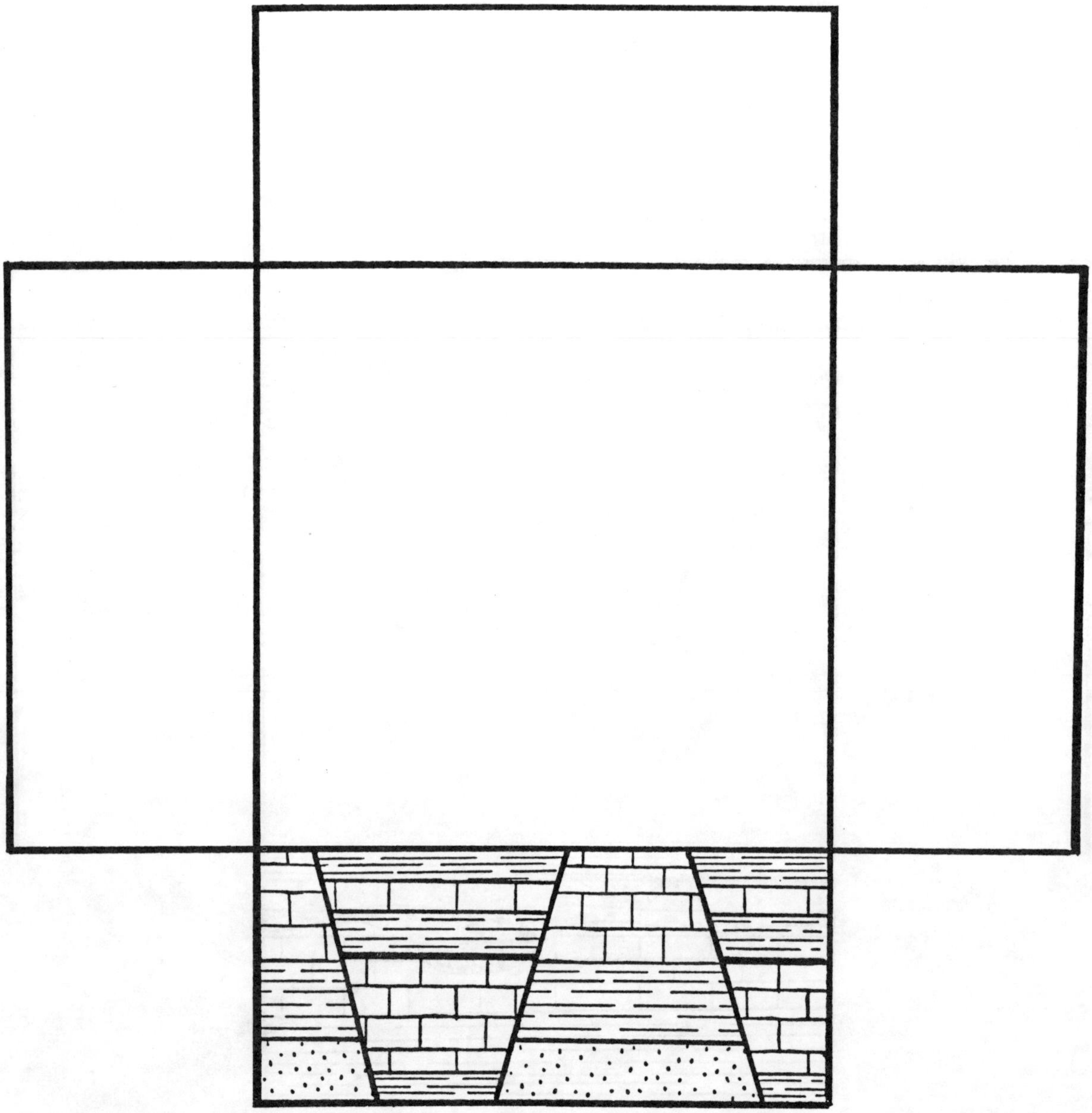

Diagram 18

Name ....
Class ....
Date ....

Worksheet 19

Diagram 19

Block Diagram Problems

Worksheet 20

Name ....

Class ....

Date ....

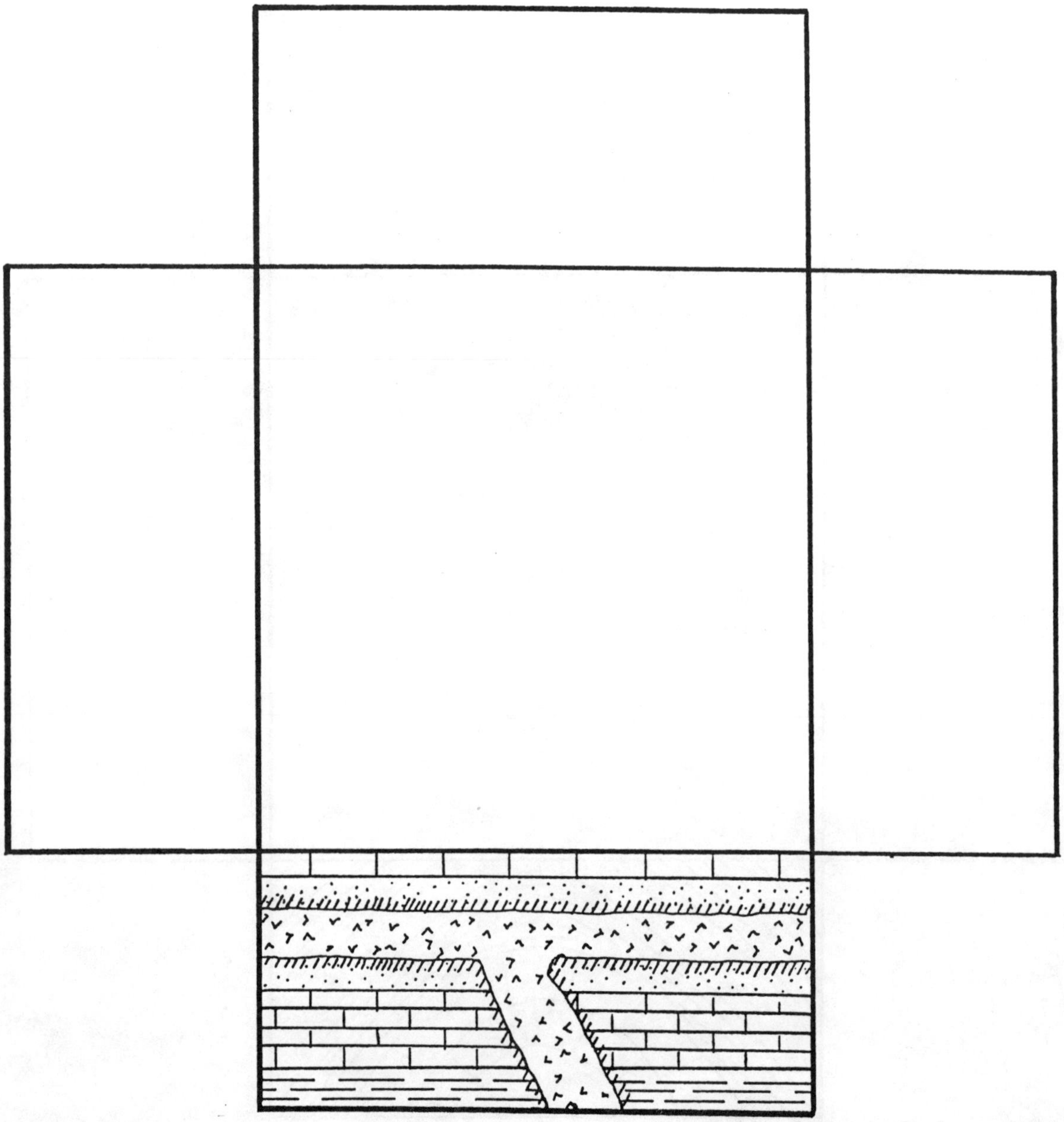

Diagram 20

Name ....

Class ....

Date ....

Worksheet 21

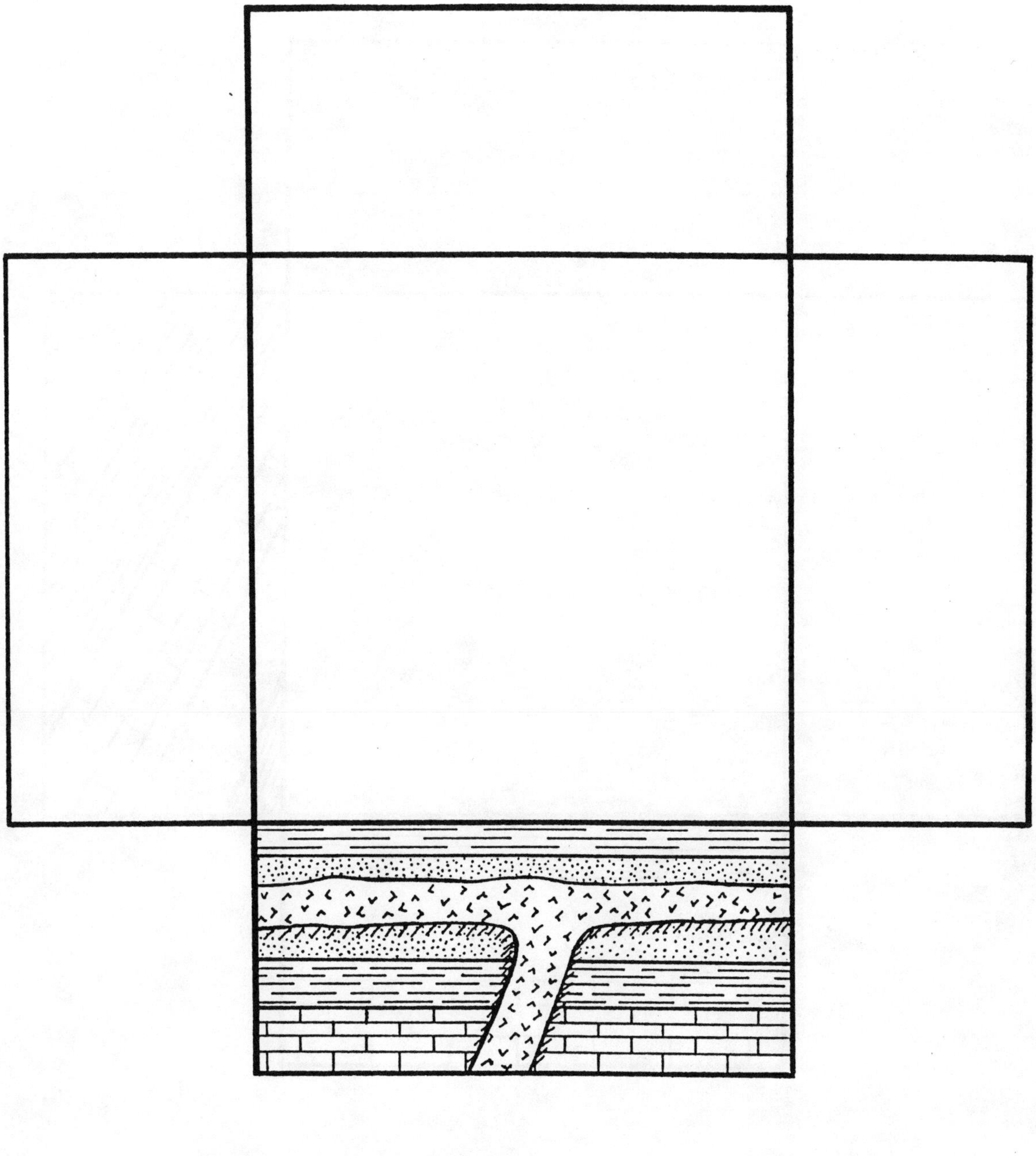

Diagram 21

Worksheet 22

Name ....
Class ....
Date ....

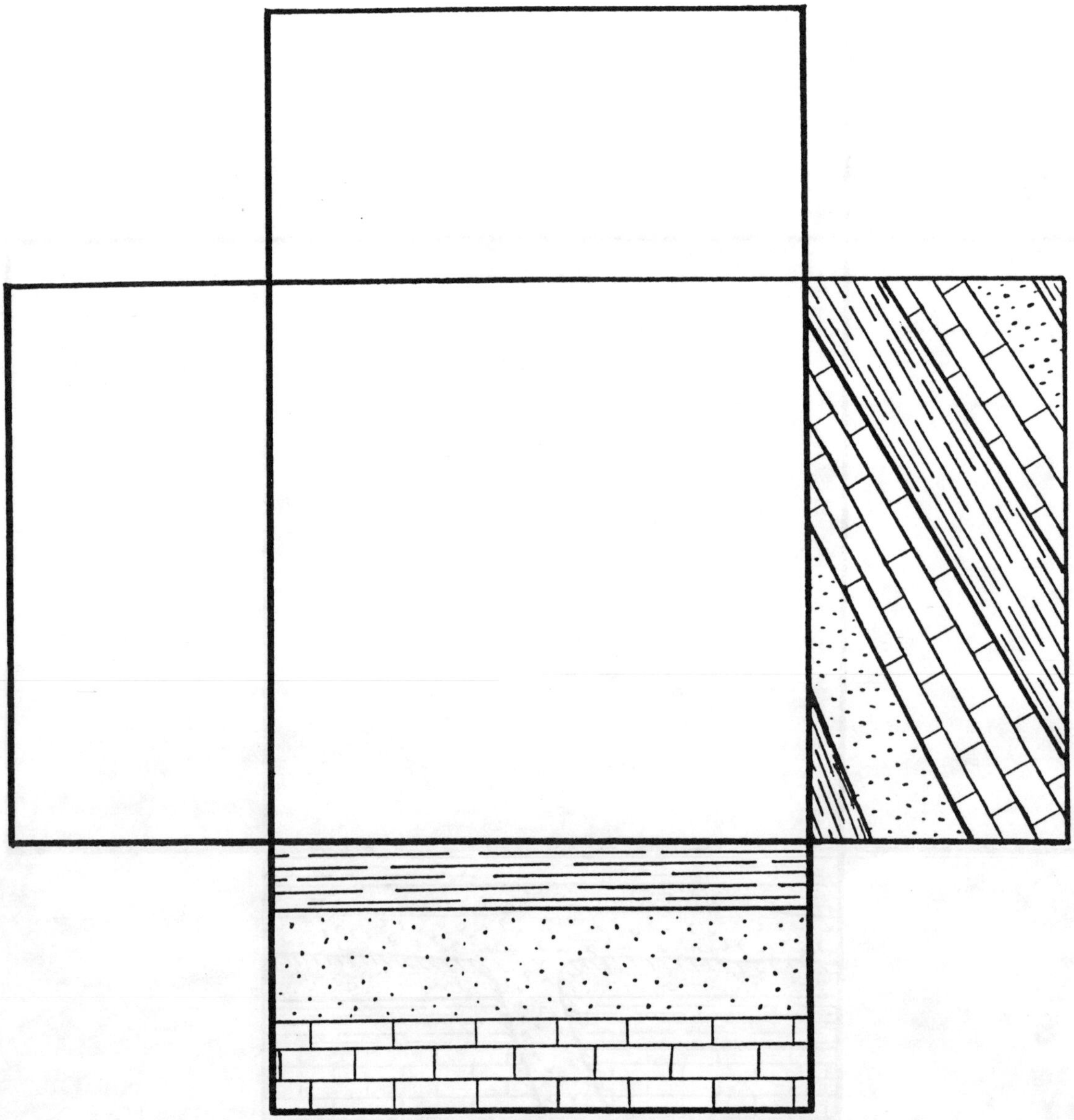

Diagram 22

Name ....

Class ....

Date ....

Worksheet 23

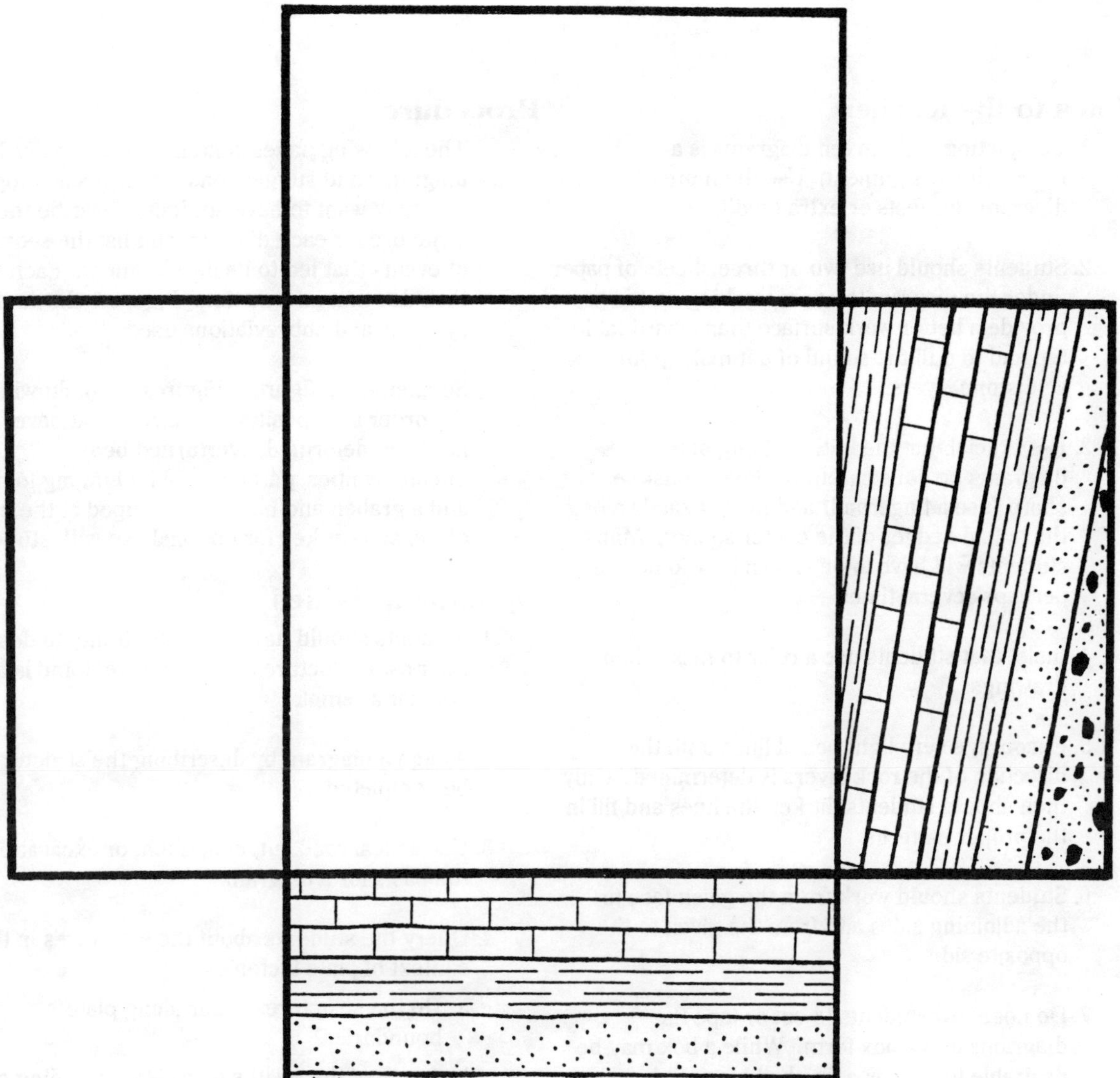

Diagram 23

Answer ✓ Key

Notes to the teacher

1. Completing eight to ten diagrams is a reasonable assignment. Use the more difficult diagrams for tests or extra credit.

2. Students should use two or three sheets of paper under the diagram they are working on to provide a better work surface than a hard table top and to dull the sound of dot-making for sandstone layers.

3. It is crucial that the initial folding of the diagrams be done carefully. Folds must be sharp (use a fingernail) and made exactly along the outside edges of the center square. Many students will have to be shown how to do this, perhaps several times.

4. Insist that students use a ruler to make their drawings.

5. Encourage very light pencil lines until the direction of the rock layers is determined. Only then should students darken the lines and fill in the proper patterns.

6. Students should work from the given face to the adjoining sides and from the given to the opposite side.

7. Do not allow students to cut or tape the diagrams into a box form. While a box may be desirable to display a finished block, it hinders the solution process. Boxed diagrams are impossible to use in the correction process.

8. Use several magazines or a stack of paper to demonstrate a syncline (Diagram 4) and anticline (Diagram 5).

Procedure

The following pages contain the 23 answer key diagrams and suggestions for discussion topics. You may want to have students describe the structure for each diagram and list the sequence of events that led to its development. Each key includes that information. Figure 1 shows the symbols and abbreviations used.

Supplemental figures (Figures 7-10) showing the order of deposition for layers that have not been deformed, overturned beds, unconformities, parts of a fold, plunging folds, and a graben and horst are grouped at the end of the answer key for optional use with students.

Optional activities

1. Students should have the opportunity to design their own structures. An intrusive dome is a popular attempt.

2. Assign a diagram by describing the structure to be completed.

3. Use a local road cut, cliff, ditch, or excavation as the basis for a diagram.

4. Query the students about the structures in the context of plate tectonics.
   a. Do the structures occur along plate boundaries?
   b. Are they the results of crustal spreading or collision (tension or compression)?
   c. Which diagrams indicate stable crustal conditions?
   d. Which diagrams show unstable conditions?

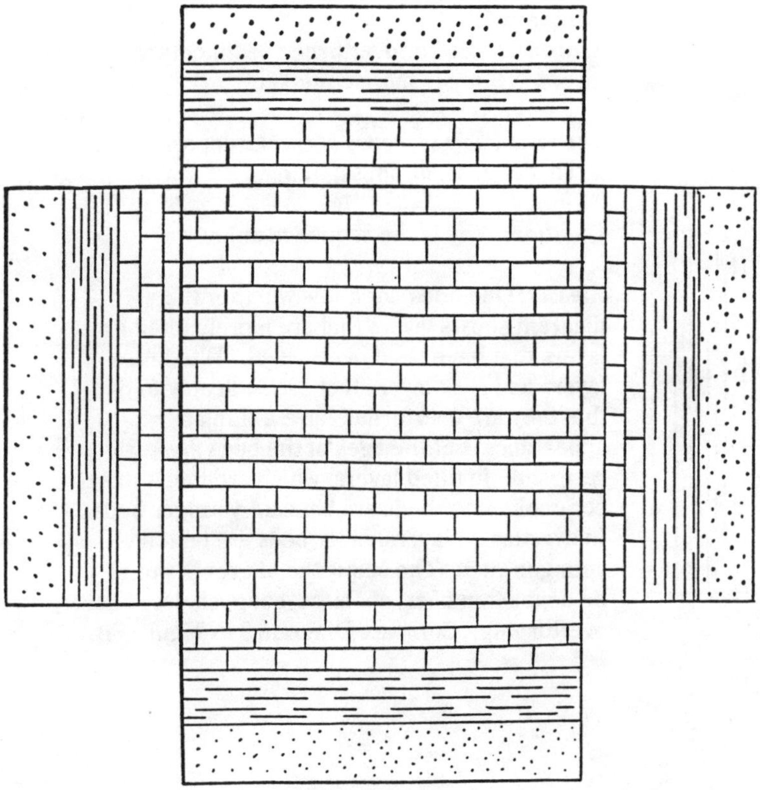

Diagram 1
Structure: flat-lying (horizontal) sedimentary beds.

Sequence of events: deposition of ss, sh, ls.

Question: Which layer is the oldest? Why?

Answer: The lowest layer of sandstone is the oldest. According to the **law of superposition,** a basic tenet of geology, in any series of sedimentary beds that have not been deformed, the oldest beds are at the bottom, and the youngest beds are at the top (Figure 7A). Some processes may overturn layers (folding) or push older rock up and over younger rocks (faulting).

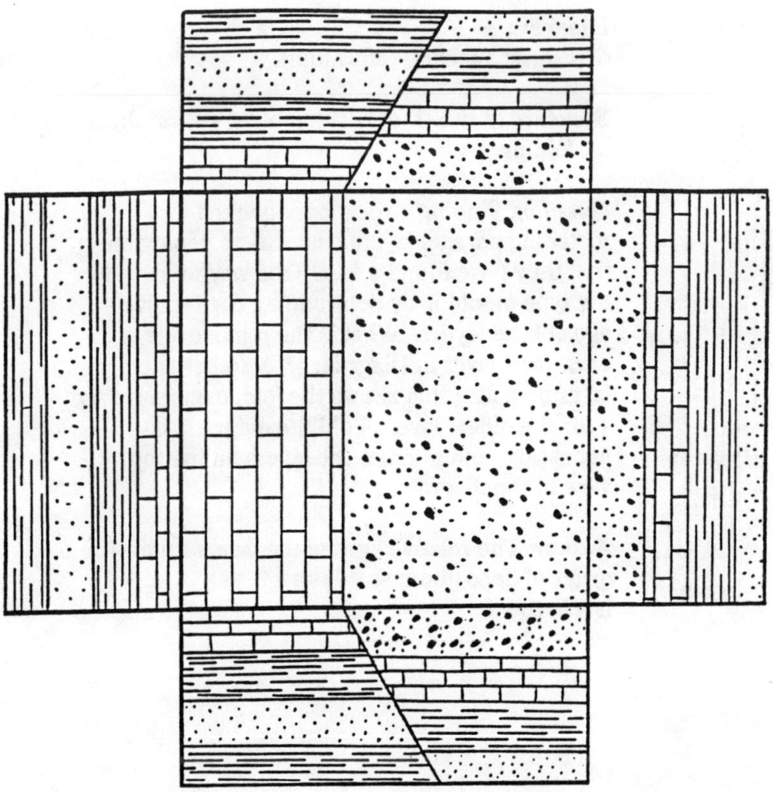

Diagram 2
Structure: faulted horizontal sedimentary layers.

Sequence of events: deposition of sh, ss, sh, ls, cong; normal faulting.

Question: What is the relative age of the fault?

Answer: The fault is younger than any of the rocks through which it cuts. By the **law of crosscutting relationships,** rocks that are folded, faulted, or invaded are older than the folding, faulting, or invading material.

Answer Key

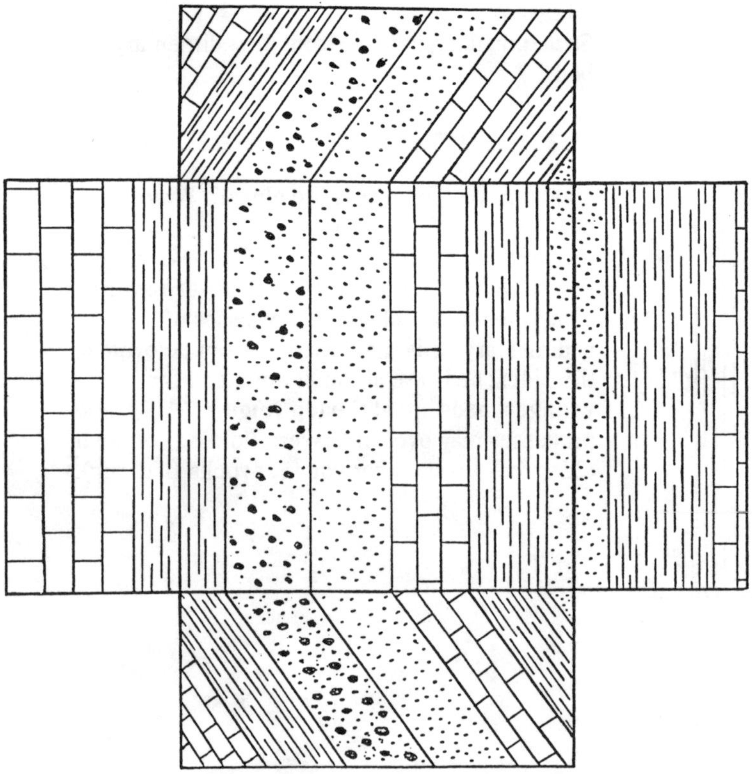

Diagram 3

Structure: tilted or overturned sedimentary layers. Tilting is more common.

The most likely sequence of events: deposition of ls, sh, cong, ss, ls, sh, ss; tilting.

Question: Why is the sequence unknown?

Answer: One does not know whether the diagram shows layers that are merely tilted or layers that have been overturned. Tilted rock layers and rock layers that are so deeply folded that they are overturned have a similar appearance, but the ages of the beds are reversed. In tilted layers, which are much more common, older beds are beneath younger beds. In overturned layers, older beds are on top of younger ones. The sequence of events would be deposition of ss, sh, ls, ss, cong, sh, ls; overturning. Compare Diagram 3 to Figure 7B.

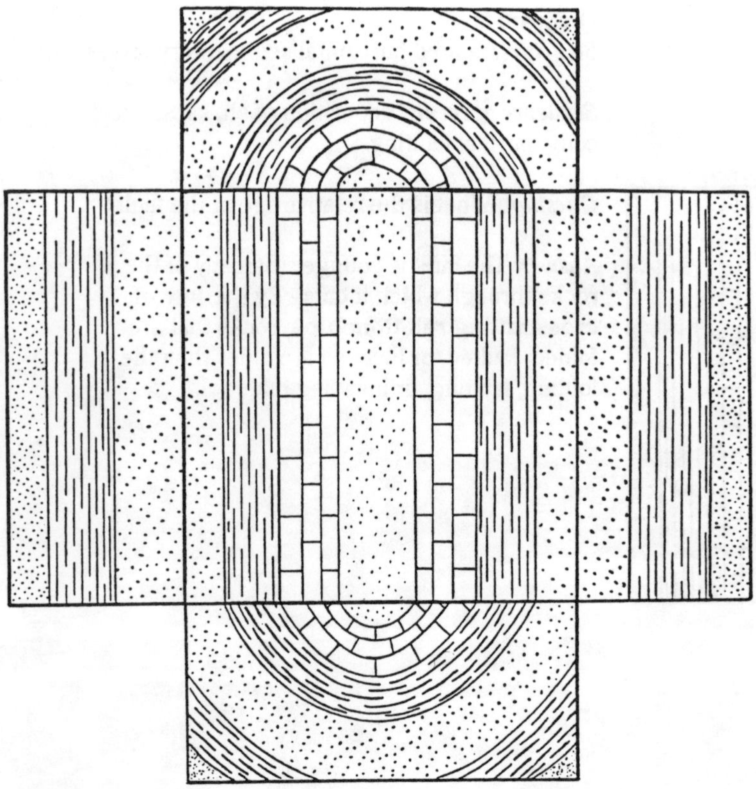

Diagram 4

Structure: syncline (downfold).

Sequence of events: deposition of ss, sh, ss, sh, ls, ss; folding.

Question: Finding a rock layer flanked by matching parallel layers suggests to geologists that they have found a fold. One way to identify the type of fold is by determining the relative ages of the layers. What is the relative age of each rock layer in Diagram 4? Number each layer on a face that shows the fold, using number 1 for the oldest layer. Fold the diagram, if necessary, and number the layers on the top view of the diagram.

Answer: The relative ages of the layers follow the order of deposition: ss = 1, sh = 2, ss = 3, sh = 4, ls = 5, ss = 6.

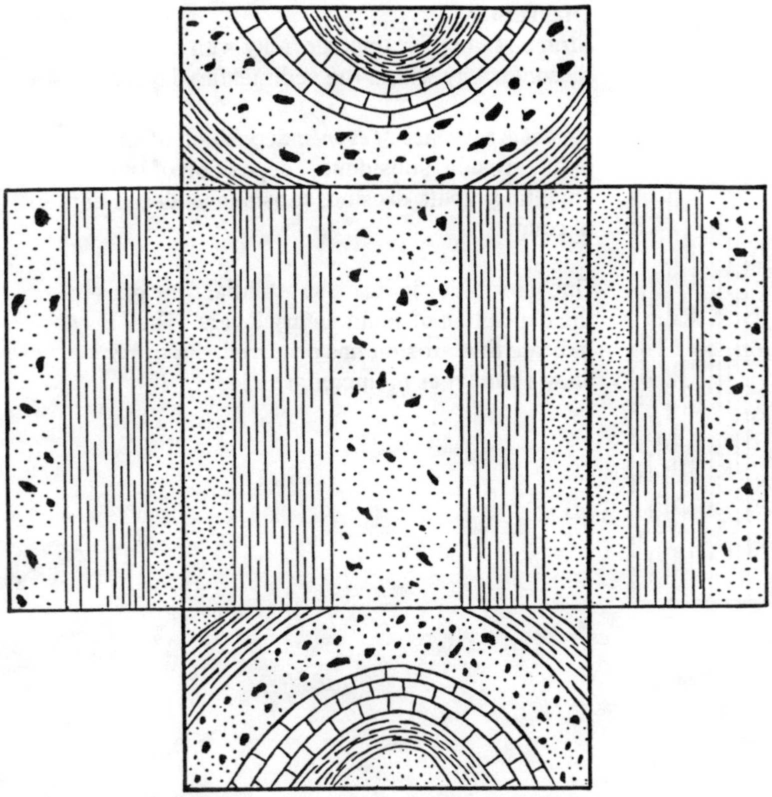

Diagram 5

Structure: anticline (upfold).

Sequence of events: deposition of ss, sh, ls, cong, sh, ss; folding.

Question: What is the relative age of each rock layer?

Answer: The relative age of each rock layer in order from oldest to youngest is ss = 1, sh = 2, ls = 3, cong = 4, sh = 5, ss = 6.

Question: What can you conclude about the relative age of rock layers as you move away from the center of a syncline? Of an anticline?

Answer: If the beds become older moving away from a center layer, the structure is a syncline. If the beds become younger moving away from a center layer, the structure is an anticline.

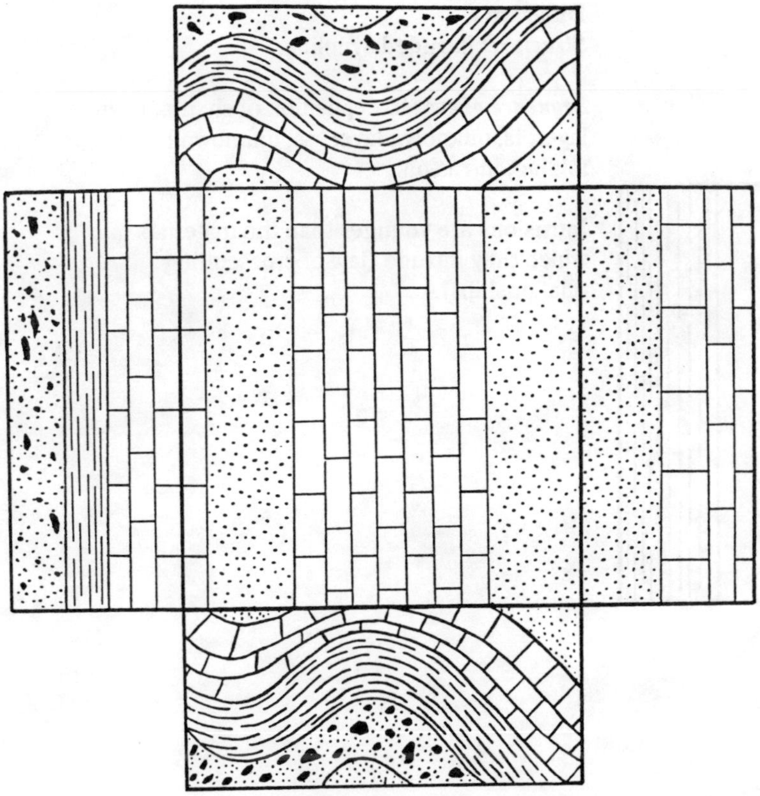

Diagram 6

Structure: synclinal and anticlinal folding.

Sequence of events: deposition of ss, cong, sh, ls, ss; folding.

A combination of an anticline and syncline is more common than an isolated syncline or anticline.

Answer✓Key

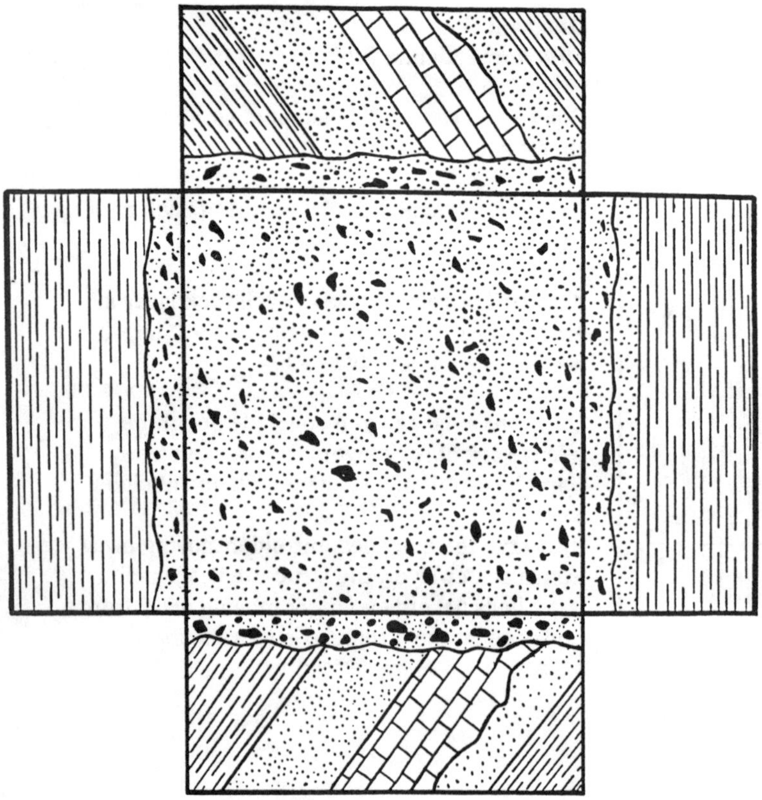

Diagram 7
Structure: angular unconformity with a disconformity within the sedimentary layers.

Probable sequence of events: deposition of sh, ss; uplift, erosion, subsidence; deposition of ls, ss, sh; tilting; uplift, erosion, subsidence; deposition of cong.

Unconformities indicate substantial time gaps in the deposition of sedimentary layers. They may indicate that erosion has occurred. Figure 8 shows two types of unconformities.

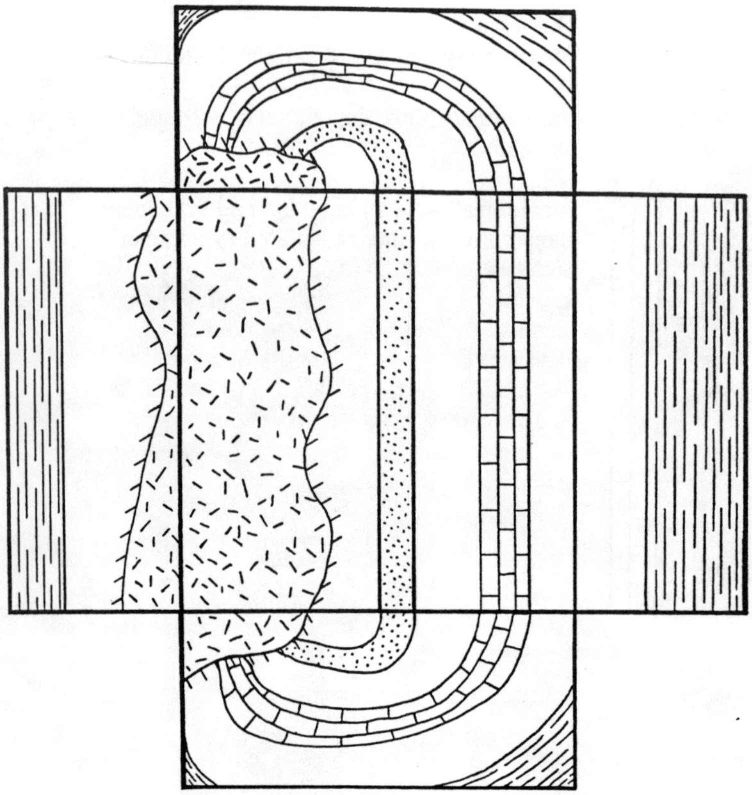

Diagram 8
Structure: intruded syncline.

Sequence of events: deposition of sh, unknown layer, ls, unknown layer, ss, unknown layer; folding; intrusion.

Intrusions are younger than the materials into which they intrude (law of crosscutting relationships).

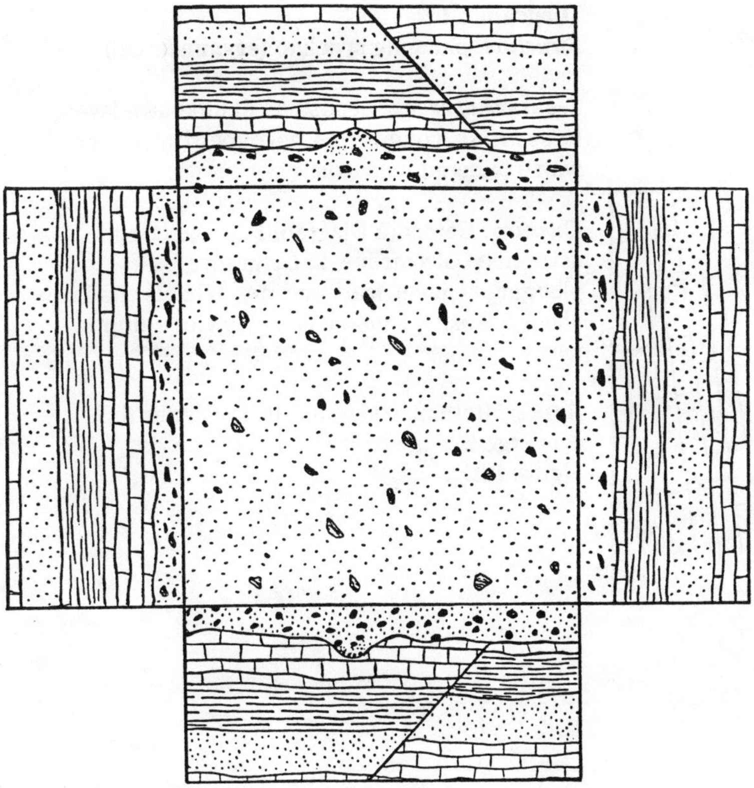

Diagram 9
Structure: faulted sedimentary beds with a disconformity showing a stream bed.

Sequence of events: deposition of ls, ss, sh, ls; normal faulting; uplift, erosion, subsidence; deposition of cong.

Question: How do you know that the conglomerate is younger than the fault?

Answer: The fault does not go through the conglomerate; therefore, that layer must not have been in place when the faulting occurred (law of crosscutting relationships). The conglomerate is younger than the fault.

Question: Were there ever more layers above the limestone?

Answer: We cannot tell if more layers were deposited above the limestone. There is no evidence of what or how much might have been present.

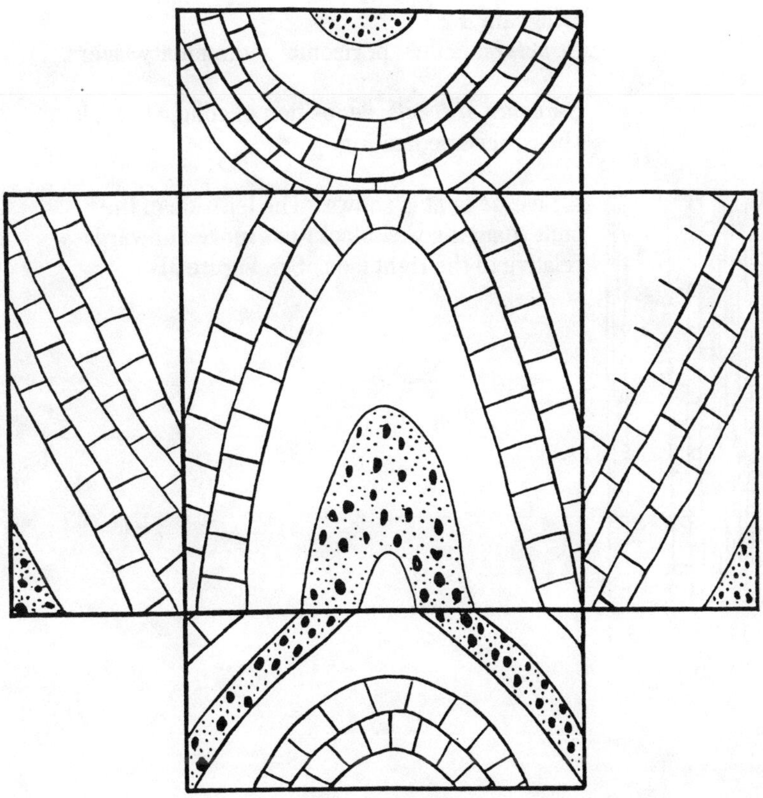

Diagram 10
Structure: plunging anticline (symmetrical).

Sequence of events: deposition of unknown layer, ls, unknown layer, cong, unknown layer, ls; symmetrical folding with a dipping or plunging fold axis.

Compare this diagram to Diagram 5. Both show anticlines. Diagram 5 shows a nonplunging anticline (axis of fold is horizontal), and Diagram 10 shows a plunging anticline (axis of fold is not horizontal). Figure 10 shows plunging folds as well as symmetrical and asymmetrical folds. Preparing a clay model of this structure is well worth the time and effort.

Answer ✓ Key

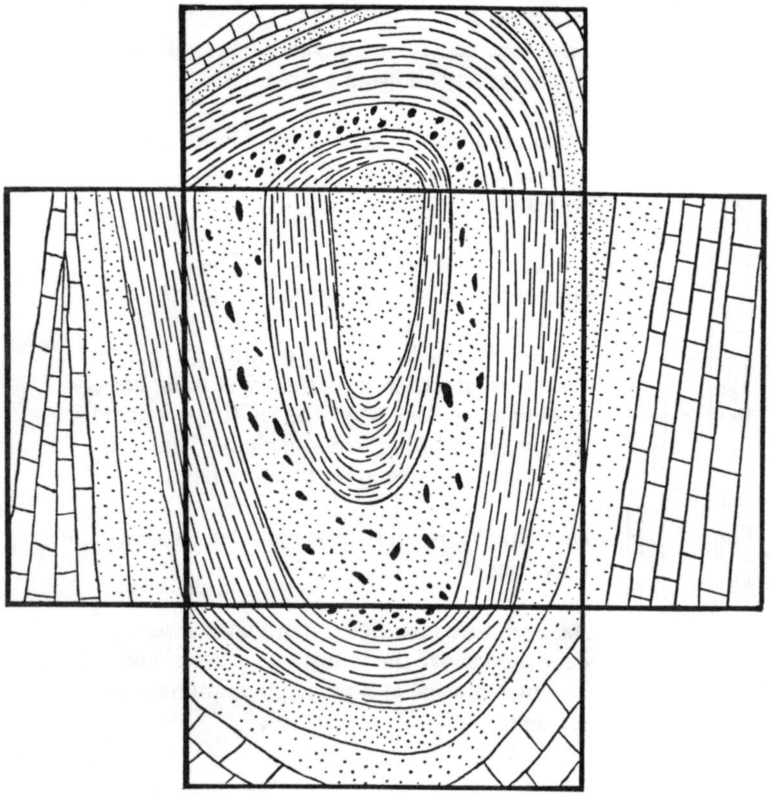

Diagram 11
Structure: plunging syncline (asymmetrical).

Sequence of events: deposition of unknown layer, ls, ss, ss, sh, cong, sh, ss; asymmetrical folding with a plunging fold axis.

Question: Compare Diagram 11, a plunging asymmetrical syncline, to Diagram 10, a plunging symmetrical anticline. What is similar about the pattern on the two map views (top face)?

Answer: Both map views show fold limbs that converge like the bow of a canoe (see also Figure 10D).

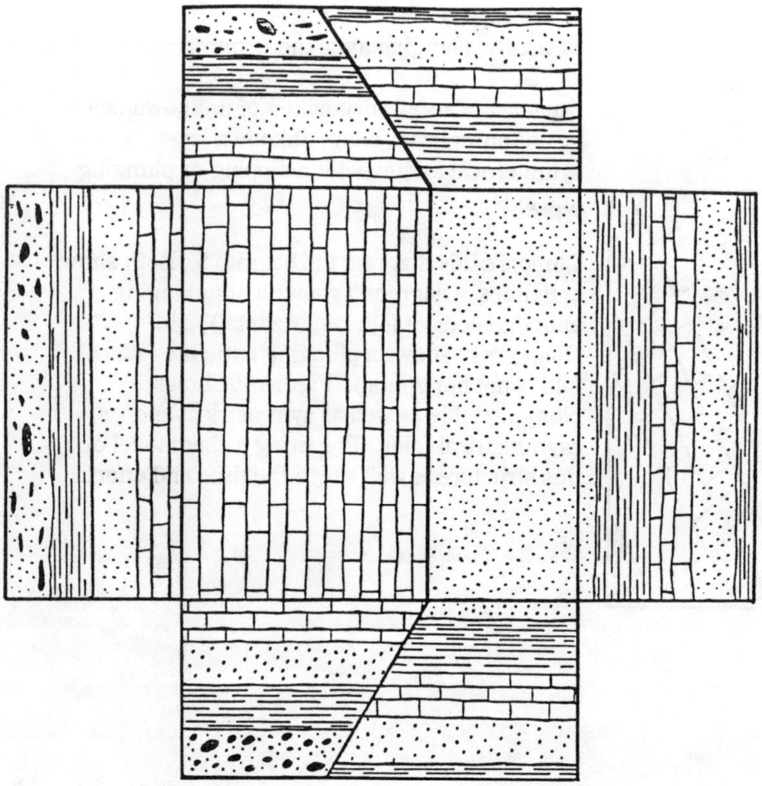

Diagram 12
Structure: faulted horizontal sedimentary layers.

Sequence of events: deposition of cong, sh, ss, ls, sh, ss; reverse faulting.

A reverse fault is shown. The left side of the fault (hanging-wall block) has moved upward relative to the right side. See Figure 3B.

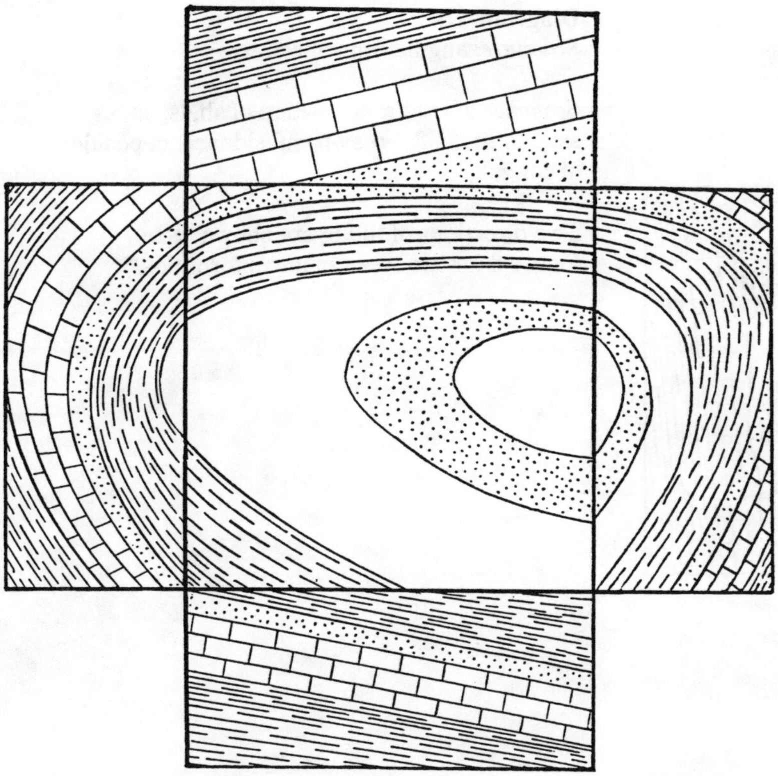

Diagram 13
Structure: plunging syncline (asymmetrical). Similar to Diagram 11.

Sequence of events: deposition of sh, ls, ss, sh, unknown layer, ss, unknown layer; asymmetrical folding with a plunging fold axis.

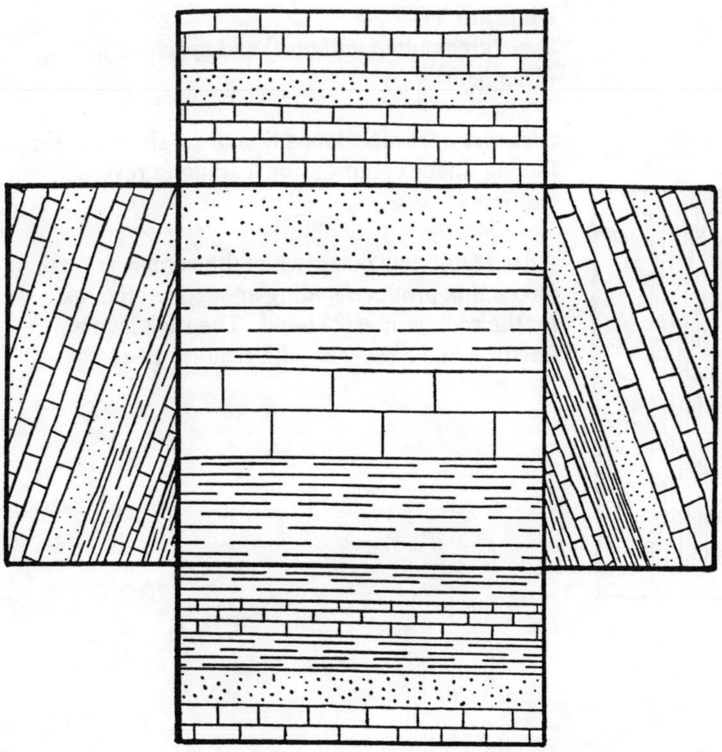

Diagram 14
Structure: tilted sedimentary layers.

Sequence of events: deposition of ls, ss, ls, ss, sh, ls, sh; tilting.

Diagrams 3, 14, 22, and 23 all show tilted sedimentary layers. They differ only in which sides are presented as the given faces.

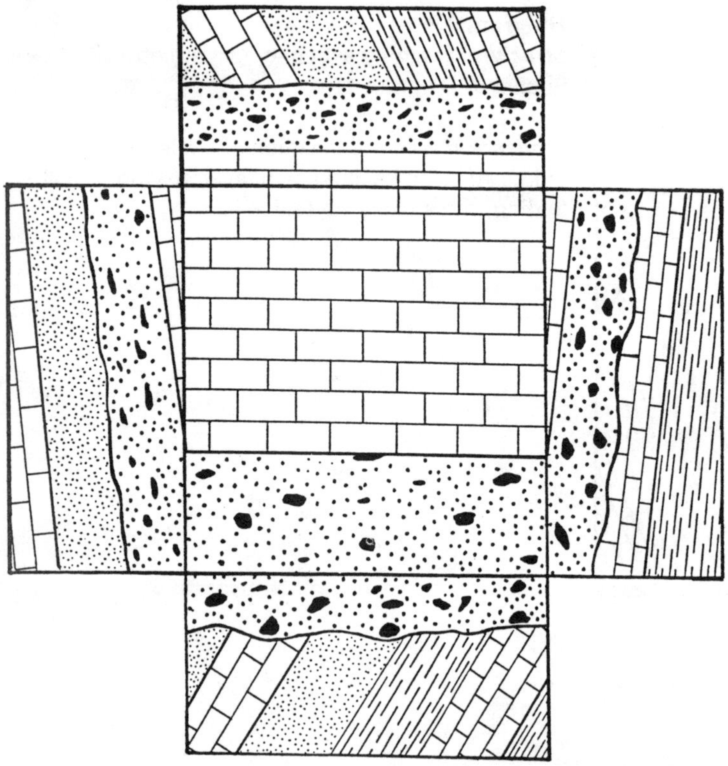

Diagram 15
Structure: angular unconformity.

Sequence of events: deposition of sh, ls, sh, ss, ls, ss; tilting; uplift, erosion, subsidence; deposition of cong, ls.

Two directions of tilting are shown. See Figure 8 for two types of unconformities.

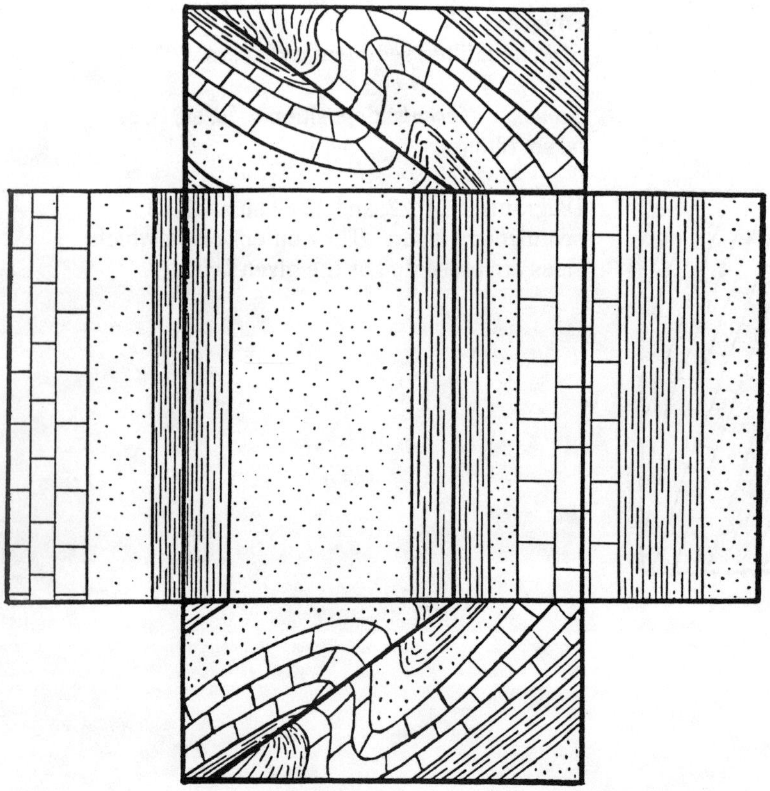

Diagram 16
Structure: faulted anticlinal and synclinal folding (drag fault).

Sequence of events: deposition of ss, sh, ls, ss, sh; folding under compression leading to reverse faulting.

A drag fault may occur when the compressional forces that produce folding are applied too fast for the rock to merely bend. The rock layers fracture, and a reverse fault results. See Figure 3B.

Answer Key

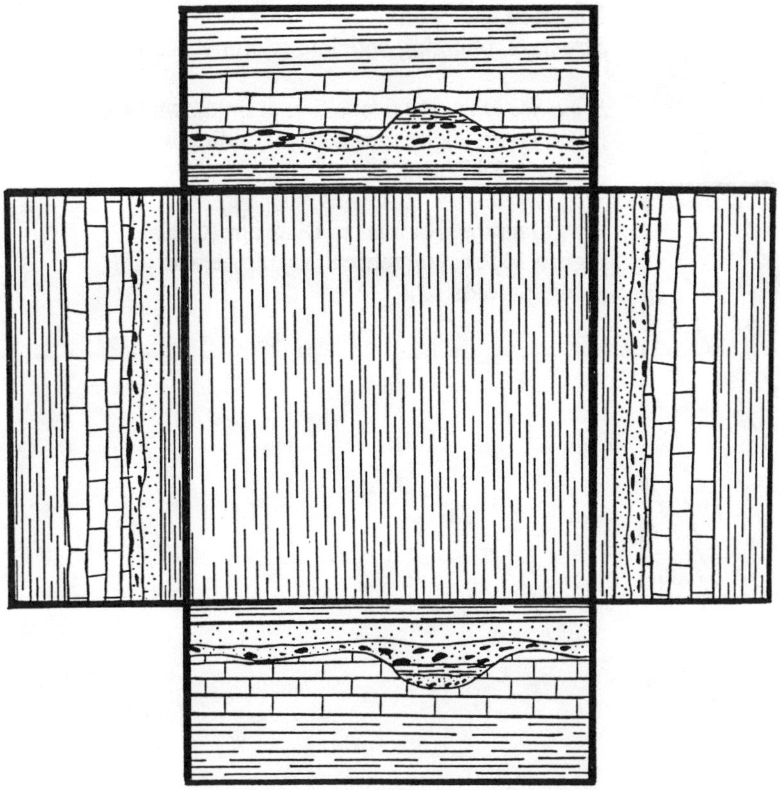

Diagram 17
Structure: horizontal sedimentary layers with a disconformity showing a stream bed.

Sequence of events: deposition of sh, ls; uplift, erosion with the development of a stream bed, subsidence; deposition of cong, ss, sh.

The stream bed is buried. Unconformities indicate a time gap, a discontinuity, in the sequence of deposition. See Figure 8 for two types of unconformities.

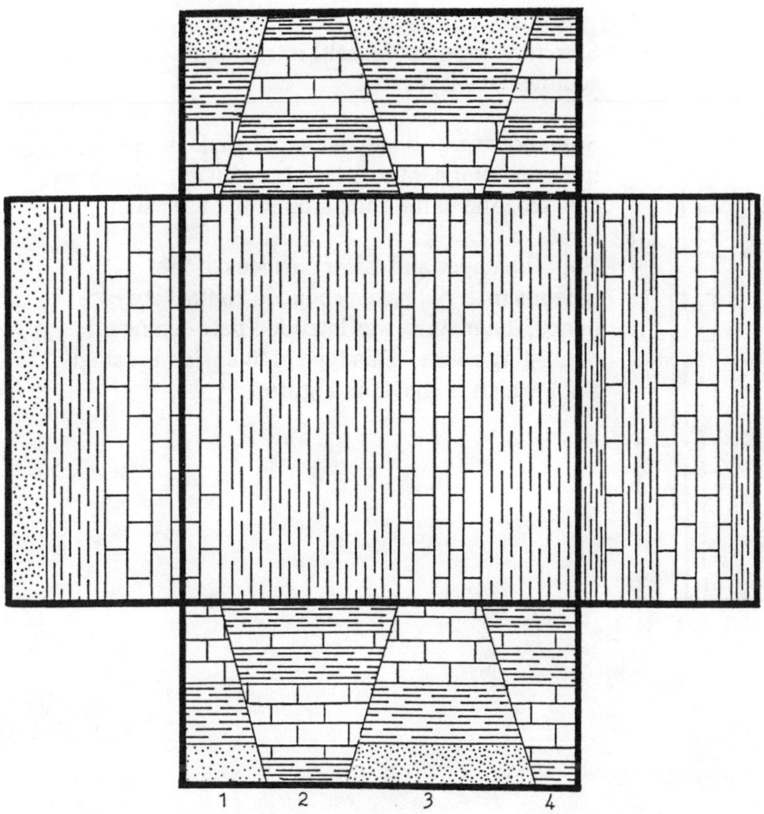

Diagram 18
Structure: normal faulting (horst and graben style).

Sequence of events: deposition of ss, sh, ls, sh, ls, sh; normal faulting caused by extension.

Note that blocks 1 and 3 have moved upward relative to blocks 2 and 4. This relative motion shows normal faulting. Figure 3 shows normal and reverse faults. Figure 9 shows a horst and a graben.

Diagram 19
Structure: plunging anticline (symmetrical).

Sequence of events: deposition of unknown layer, ls, sh, cong, sh, ls; symmetrical folding with a plunging fold axis.

This diagram is similar to Diagram 10. See Figure 10 for symmetrical, asymmetrical, and plunging folds.

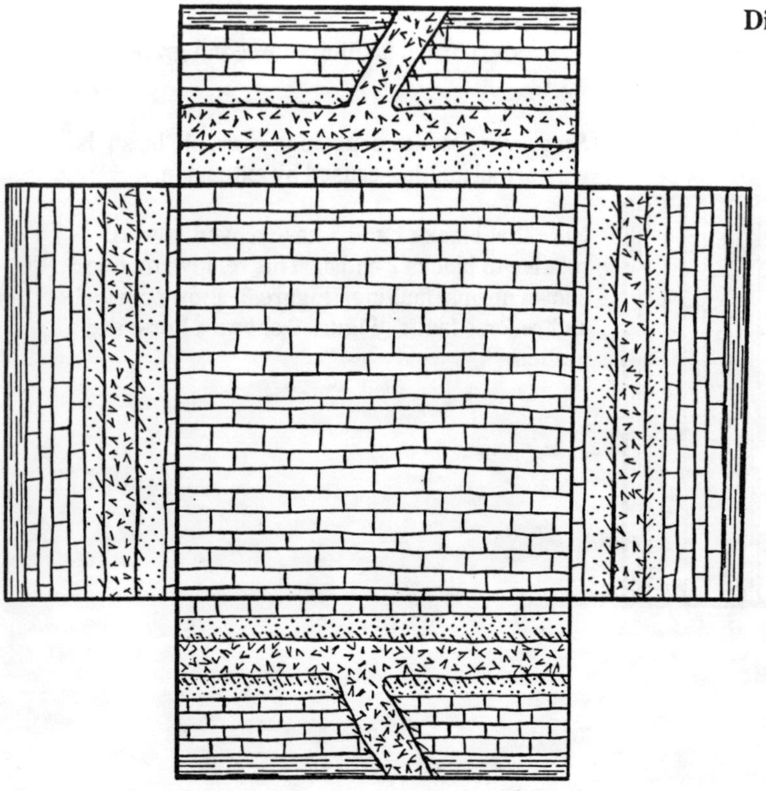

Diagram 20
Structure: igneous (basalt) intrusion into sedimentary layers.

Sequence of events: deposition of sh, ls, ss, ss, ls; intrusion of igneous material producing a contact zone in the rock around the intrusion.

Both the upper and lower surfaces of the igneous material have produced contact zones; hence, the magma was intruded between layers of existing rocks. This type of intrusion is called a sill, which is shown in Figure 5.

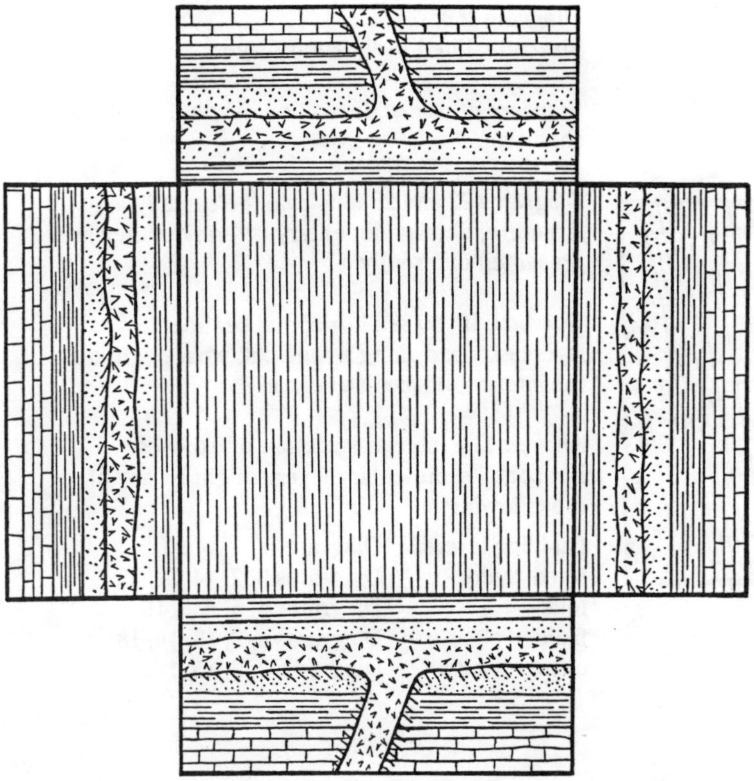

Diagram 21
Structure: buried lava flow.

Sequence of events: deposition of ls, sh, ss; lava flow; deposition of ss, sh. Uplift could occur after the deposition of ss and before the lava flow; subsidence could occur after the lava flow.

Only one contact zone is observed, indicating no material was above the lava when it was emplaced. The lava flow was subsequently buried.

Question: Did this lava flow occur under water or on the surface of the land?

Answer: One cannot tell from the information provided. Pillow lava would indicate an underwater flow.

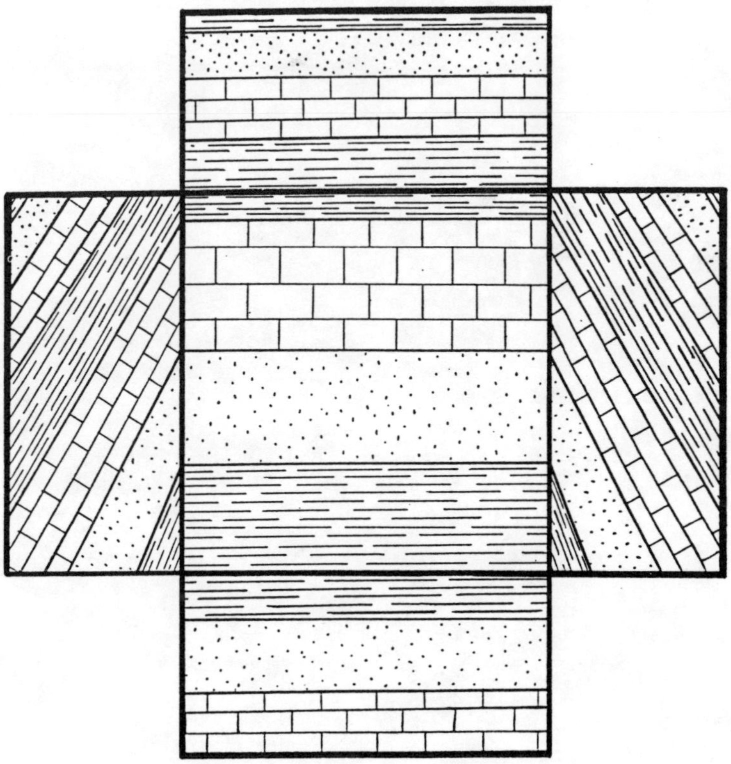

Diagram 22
Structure: tilted sedimentary layers.

Sequence of events: deposition of sh, ss, ls, sh, ls, ss, sh; tilting.

Answer Key

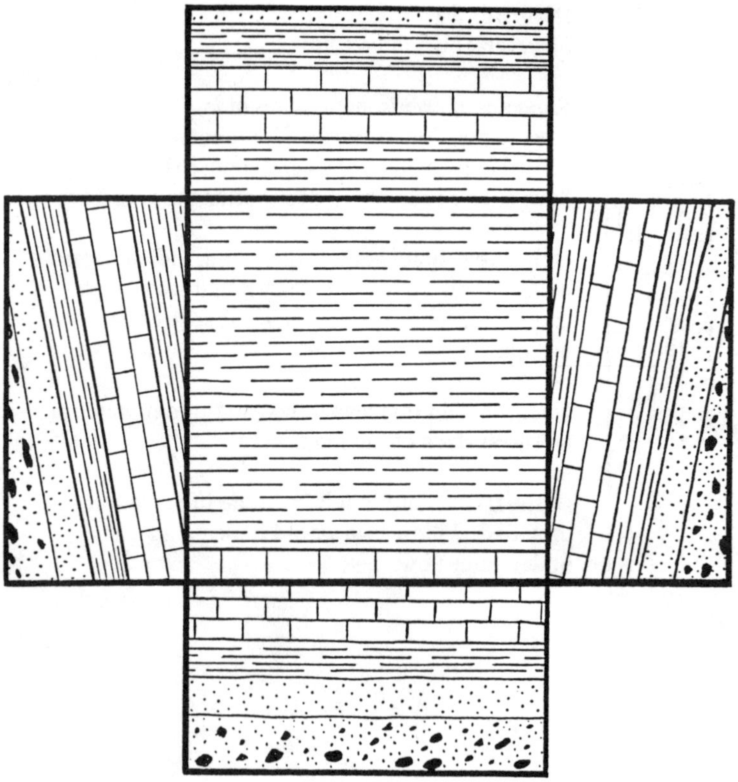

Diagram 23
Structure: tilted sedimentary layers.

Sequence of events: deposition of cong, ss, sh, ls, sh; tilting.

Diagrams 22 and 23 illustrate the same type of structure, tilted sedimentary layers, but they are presented differently.

Question: If you see horizontal sedimentary layers exposed along a road cut, what might be the interior, unseen structure?

Answer: The interior structure of exposed horizontal sedimentary layers might be any of the following:

flat-lying layers:	Diagram 1
tilted layers:	Diagrams 3, 7, 14, 22, 23
fold:	Diagrams 4, 5, 6, 8, 16
fault:	Diagrams 2, 9, 12, 16, 18
unconformity:	Diagrams 7, 9, 15, 17

Supplemental figures

Figure 7. Relative ages of rock layers. (A) Order of deposition when rock layers have not been deformed: younger beds overlie older beds. Conglomerate (1) is the oldest bed, and shale (3) is the youngest. (B) Overturned rock layers showing how the position of the beds becomes reversed from that in which they were originally deposited. The oldest bed is on top. Area in square is section that is exposed.

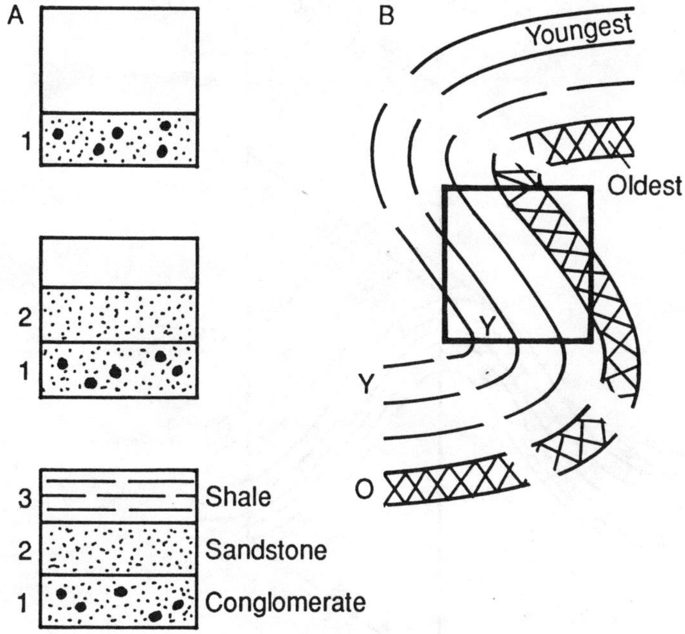

Figure 8. Two types of unconformities. (A) Angular unconformity, rock layers above and below the unconformity meet at an angle. (B) Disconformity, rock layers above and below the unconformity are parallel.

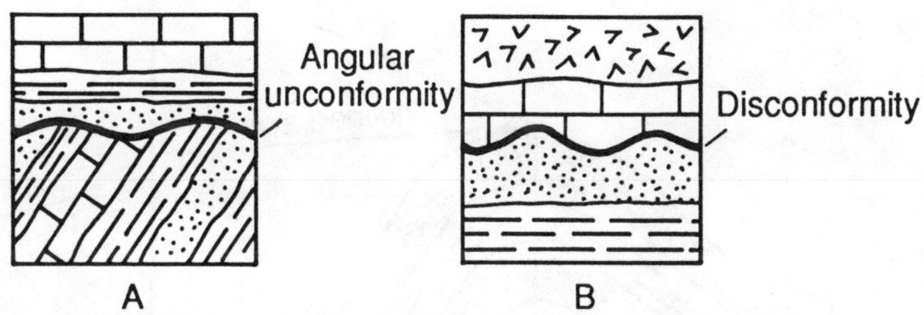

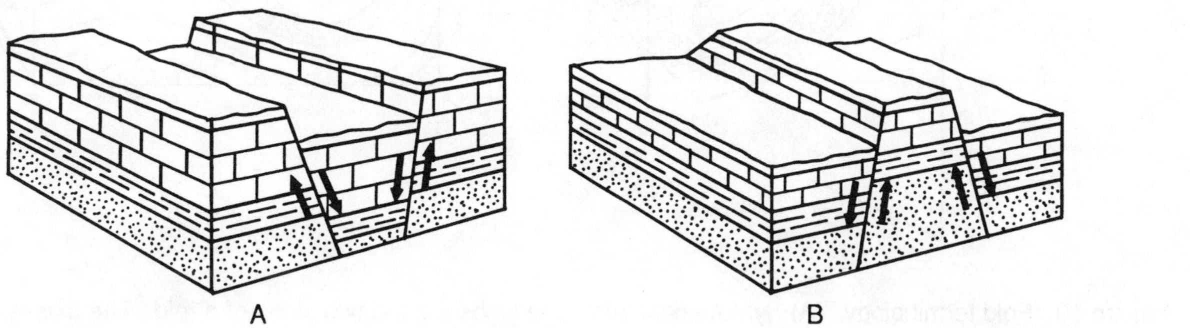

Figure 9. (A) Graben, a trench-like structure bounded by two parallel normal faults. It is formed when crustal tensions cause a fault-bounded block to move downward. (B) Horst, a ridge bounded by parallel normal and sometimes reverse faults. It is formed when crustal tensions cause a fault-bounded block to move upward.

Supplemental figures

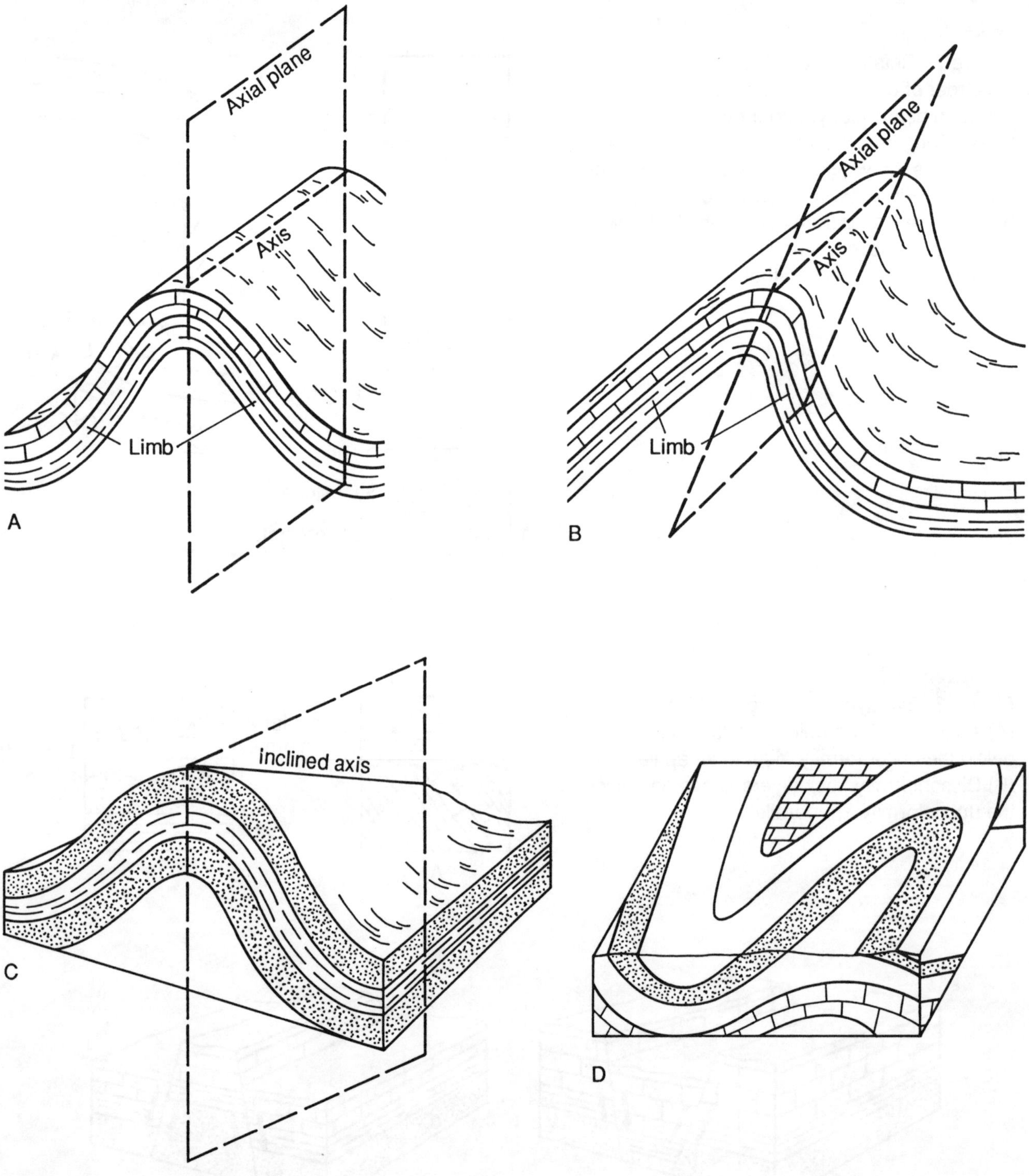

Figure 10. Fold terminology. (A) Symmetrical fold. The limbs are the two sides of a fold. The axis is the median line between the limbs. The axial plane is the surface that passes through the axis of each rock layer. The fold is symmetrical: the beds in each limb dip equally on either side of the axial plane. (B) Asymmetrical fold. The beds in one limb dip more steeply than those in the other. (C) Plunging fold. The axis is inclined, not horizontal. Compare the inclination of the axis of a plunging fold (C) to that of a nonplunging fold (A). (D) Plunging fold. The map (top) view of a plunging fold shows limbs that converge like the bow of a canoe.

What Earthquake Waves Tell Us about Earth's Interior

Donald G. Korba
Birchmount Park Collegiate Institute
Scarborough, Ontario, Canada

Objective

To determine graphically the boundary between the mantle and outer core and the boundary between the inner and outer core.

Time required

Two class periods

Materials

per student
Protractor: 10 cm (4 in.) or larger

Background

Note: Students should know how to use a protractor to measure angular distance.

When an earthquake occurs, vibrations called **seismic waves** are produced. These waves travel outward from an earthquake's **focus**, its point of origin, in all directions. Two kinds of seismic waves travel through the Earth: **P-waves** and **S-waves.** The third kind, surface waves, results when P- and S-waves reach the surface and travel outward along the surface from an earthquake's **epicenter**, the point on the surface directly above its focus. By studying P- and S-waves, scientists have been able to develop a model of the Earth's interior.

P-waves can travel through solids, liquids, and gases, and they move faster through dense material than through less dense material. Studies of earthquake waves have shown that P-wave velocity increases or decreases at certain depths inside the Earth. A change in velocity suggests the presence of a boundary between materials of different densities.

The outer layer of the Earth is known as the **crust.** As P-waves travel from an epicenter through the crust, they suddenly increase in speed at a depth of about 35 km. The change in speed indicates they are entering a material of higher density than the crust. This layer is called the **mantle.** At a depth of about 2,900 km, P-waves sharply decrease in speed, indicating they are entering the core, a material of lower density than the mantle. At about 5,100 km, P-waves speed up, signifying the presence of a denser material. Scientists determined that the core can be divided into an **outer core** and an **inner core**, with the inner core having denser material.

S-waves can travel through solid material only, not through liquids. Like that of P-waves, their speed changes according to the density of the material through which they are traveling, speeding up in dense material and slowing down in less dense material.

Scientists have found that S-waves cannot be detected on the side of the Earth opposite an earthquake; therefore S-waves were not able to travel through the Earth's core. Since S-waves cannot travel through liquids, and P-wave velocity suggested that the outer core was less dense than the inner core, scientists think that the outer core may be somewhat fluid and the inner core solid.

Shadow zone: P- and S-waves are not detected at seismograph stations that are at an angular distance greater than 103° from the epicenter of an earthquake. P-waves are detected at an angular distance of about 143° or more. These findings indicate a zone on the Earth's surface between 103° and 143° from an epicenter in which no P- or S-waves can be detected. This region is called an earthquake's **shadow zone.** Its existence can be explained by the way seismic waves bend as they travel through the Earth and by the inability of S-waves to travel through liquids.

When seismic waves travel through material that changes density gradually, they change direction gradually. When they travel from one material to another material of different density, they change direction sharply. As P- and S-waves travel through the mantle, they may bend slightly because of slightly different densities. When P-waves enter the boundary between the mantle and the core, the difference in densities causes them to bend sharply. They are reflected away from the core and reach the surface at about 103° from an epicenter, or they are bent inward and travel through the core. When they leave the core, they are bent sharply again and reach the surface at about 143° from an epicenter. When S-waves reach the boundary between the mantle and the core, they are reflected back because they cannot travel through the dense fluid in the outer core. They reach the surface at about 143° from an epicenter. The result is the shadow zone between 103° and 143° from an epicenter where no P- and S-waves are detected.

Procedure

1. Circles 1 and 2 on the worksheet represent cross sections of the Earth. Each dot shows the focus of an earthquake. On circle 1, draw and number (in any order) a dot on the surface of the Earth above each focus to show the epicenter of each earthquake.

2. Using a protractor, with the vertex at the Earth's center, mark off an angle of 103° on the circle starting 0° at epicenter 1. Repeat on the other side of epicenter 1. Then draw a line joining each 103° mark to epicenter 1 (Figure 1). You should draw only the solid lines.

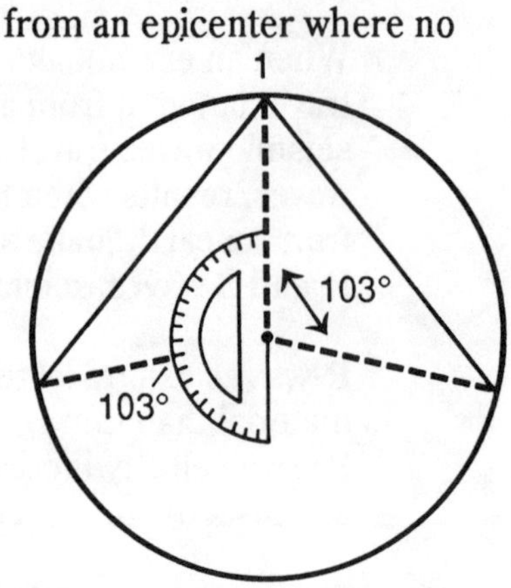

Figure 1. Marking angles from an epicenter.

3. Repeat step 2 for each of the 15 other epicenters.

4. On circle 2, repeat steps 1, 2, and 3, marking an angle of 143° instead of 103°.

Questions

1. What did the lines you drew in steps 2 and 3 seem to outline?
2. Because scientists have found that S-waves of earthquakes cease to be detected beyond about 103° from each epicenter, each line you drew in steps 2 and 3 could represent the path of the last S-wave that passes directly through the Earth. What must these last S-waves indicate about Earth's interior at this depth?
3. The zone on the Earth's surface between 103° and 143° from each epicenter is given a special name. What is its name and what waves are received beyond 143°?
4. What boundary did you discover in step 4?

Additional resources

Heller, Robert L., et al. *Earth Science.* 2d ed. New York: McGraw-Hill Book Co., 1978.

Montgomery, Carla W. *Physical Geology.* Dubuque: William C. Brown Group, 1987.

Namowitz, Samuel N., and Spaulding, Nancy E. *Earth Science.* 9th ed. Lexington: D.C. Heath and Co., 1989.

Tarbuck, Edward J., and Lutgens, Frederick K. *The Earth: An Introduction to Physical Geology.* 2d ed. Columbus: Merrill Publishing Co., 1987.

Name..........

Class..........

Date..........

Worksheet

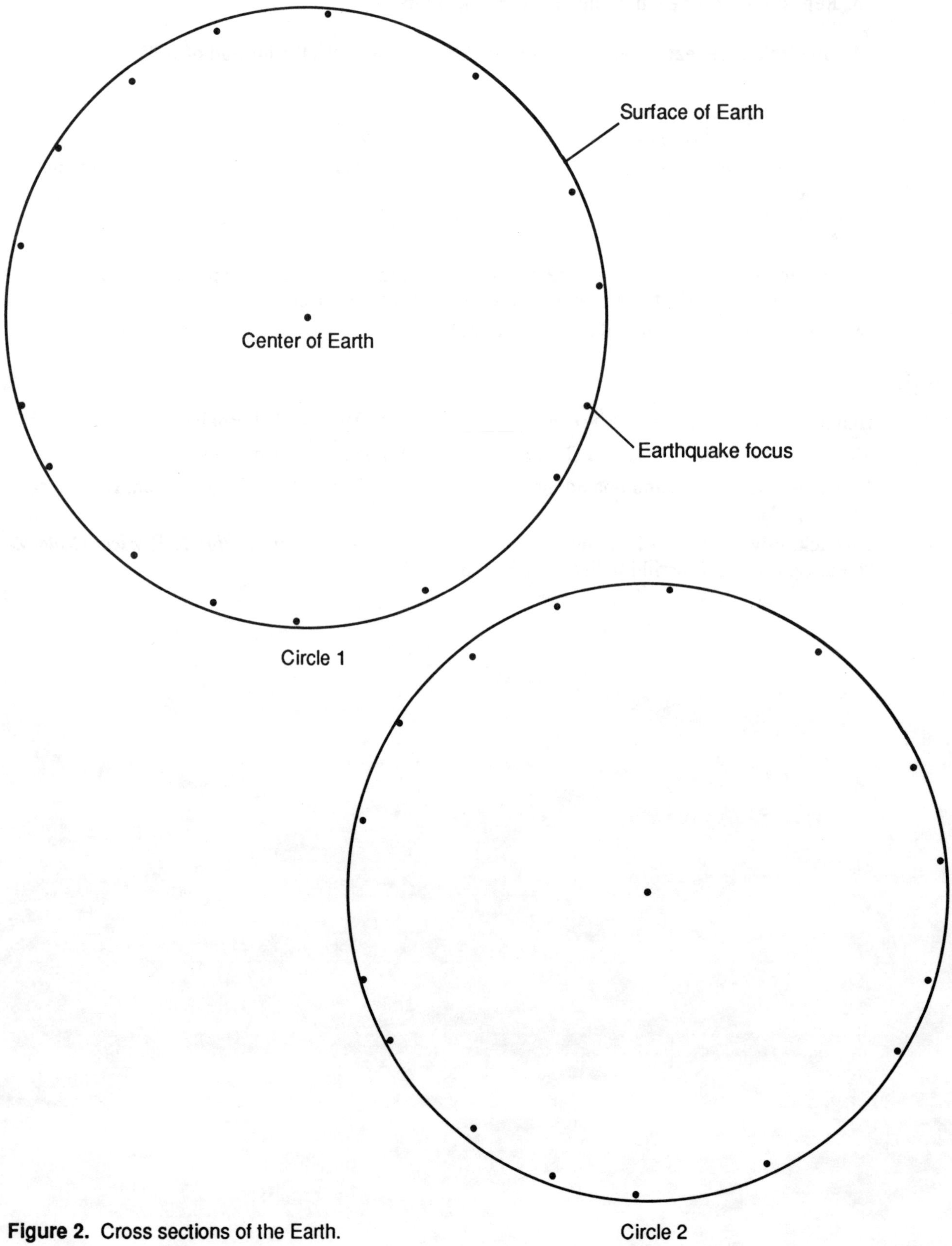

Figure 2. Cross sections of the Earth.

Answer Key

Procedure

Steps 2 and 3. The lines using 103° should appear as shown below on the left.

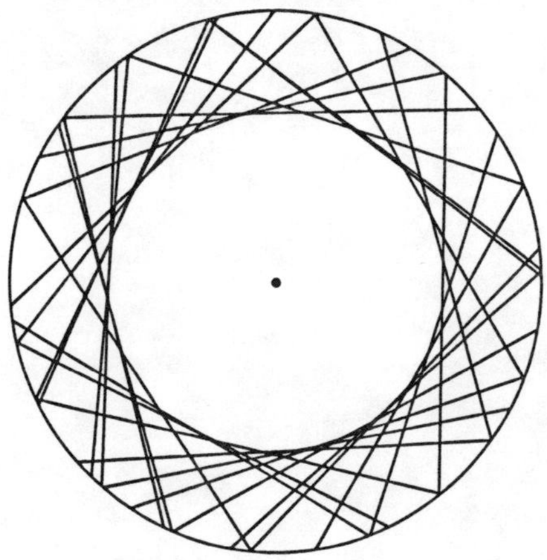

103°

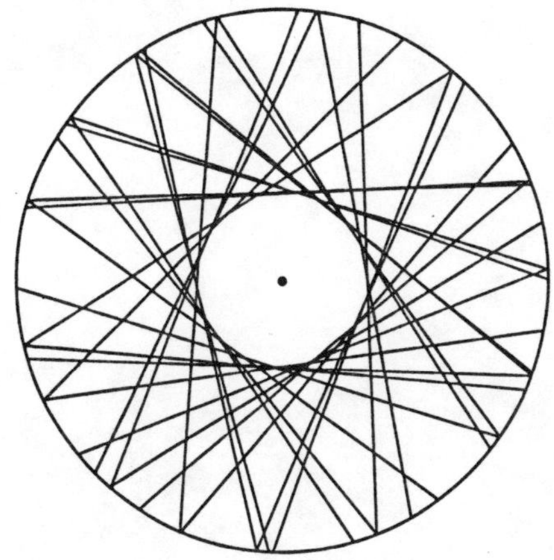

143°

Step 4. The lines using 143° should appear as shown above on the right.

Questions

1. The lines seem to outline a circle.

2. The last S-waves indicate that beyond the "circle" there may be a dense fluid within the Earth.

3. This zone is called the shadow zone. The waves received beyond 143° are P-waves only.

4. The boundary between the outer and inner core is outlined.

Epicenters of the 1964 Alaskan Earthquake

ROBERT A. CHRISTMAN

Western Washington University
Bellingham, Washington

Objectives

To learn how to locate the epicenter of an earthquake.

To study the distribution of the aftershocks of the 1964 Alaskan earthquake.

To understand how the Alaskan earthquake might be explained by the theory of plate tectonics.

Time required

Two class periods

Materials

per student
Compass (for drawing circles)

Background

Note: Students should know how to plot locations on a map using latitude and longitude.

After an earthquake has occurred, geologists need to find its exact location. How can this be done? By talking to people, it might be very difficult to decide where the ground shook the most. An earthquake might feel one way to a person standing outside on a rock and another way to a person on the third floor of a building with a weak foundation on soft sediments. Also, the place of the strongest movement might be in unpopulated areas. The major shock may occur far from towns, where witnesses are scarce.

To find the exact location, seismologists study the waves that are produced by the energy released during the earthquake. Earthquake waves travel through the ground. Their amplitudes and times of arrival may be recorded by a seismograph on a record called a seismogram. By studying seismograms, seismologists can determine the distance the waves have traveled. With this information, they can locate the earthquake's **epicenter**, the point on the Earth's surface directly above the earthquake's **focus**, its point of origin.

A seismogram (Figure 1) shows the arrival times of two different kinds of seismic waves produced by the earthquake. These waves are called **P**- (primary) and **S**- (secondary) **waves**. The P-waves travel faster than the S-waves, so they reach a seismograph station first. The time

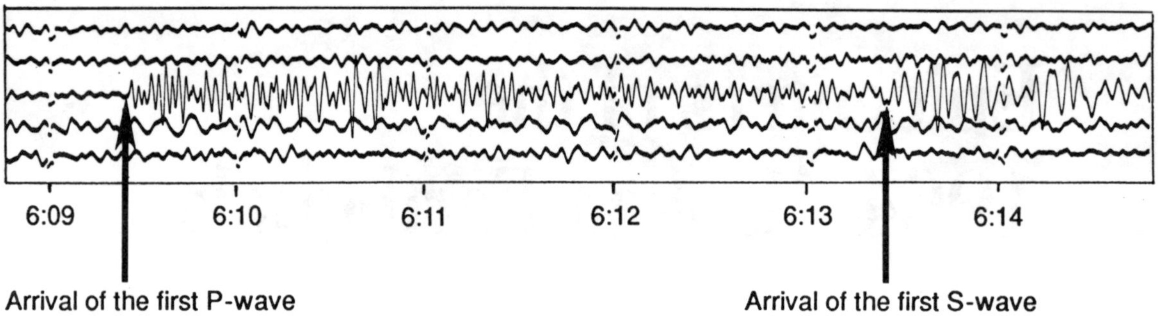

Figure 1. Section of the Bellingham station seismogram showing the record of the 6:05 Greenwich Mean Time (GMT), March 29 aftershock of the 1964 Alaskan earthquake. The first P-wave arrived at 6:09:24 GMT (hours, minutes, seconds), and the first S-wave arrived at 6:13:24 GMT for an S-P time interval of four minutes. (Time is given in 24-hour clock time. For example, 6:09 is 6:09 a.m.)

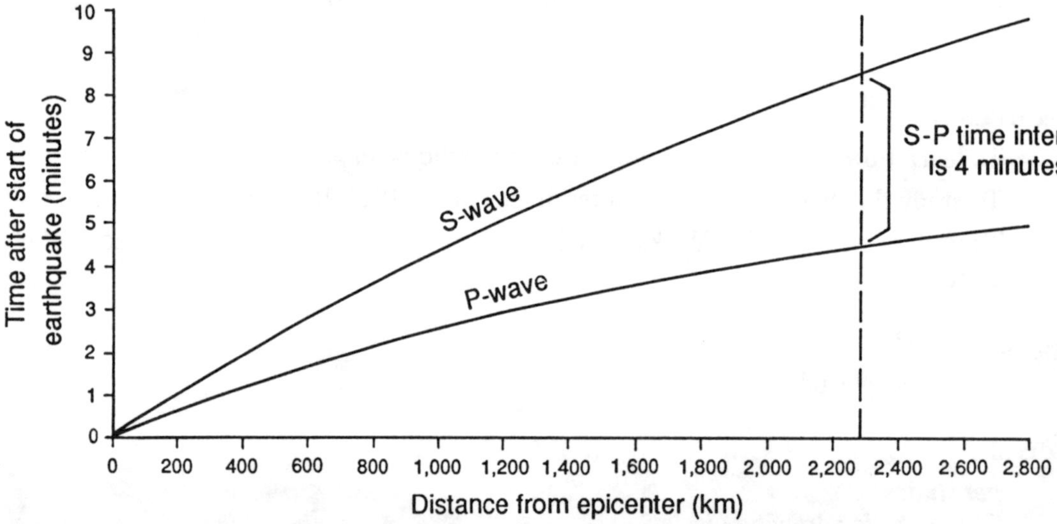

Figure 2. Travel-time graph for P- and S-waves. The S-P time interval (four minutes) indicates that the epicenter of the earthquake is about 2,280 km from the seismograph station.

interval between the arrival of the first P-wave and that of the first S-wave, called the S-P time interval, can be converted into distance using a standard travel-time graph.

Figure 2 shows a travel-time graph with the S-P time interval from the seismogram in Figure 1 marked on it. The bracket indicates the area where the travel times for the two waves are four minutes apart. Reading the value from the x-axis shows that the seismograph station is about 2,280 km from the earthquake's epicenter. The farther the waves travel, the longer the S-P time interval.

Using the distance determined from the travel-time graph as a radius, a circle can be drawn on a map around the seismograph station. The epicenter will be located somewhere along this circle.

Using data from another seismograph station, located elsewhere, seismologists can determine its distance from the earthquake. Another circle can be plotted on the map, and the intersections of the two circles will show two possible locations for the earthquake. If information from a third station is added to the map, the intersections of the three circles will show the approximate position of the earthquake.

Seismograph records of P- and S-waves help determine not only an earthquake's epicenter but also its magnitude, the amount of energy released at its focus. The amplitude or height of the waves recorded is related to an earthquake's magnitude. Charles Richter, an American seismologist, developed a magnitude scale, the Richter scale, which uses numbers to show levels of magnitude. Starting with one, each successive number indicates an earthquake having 10 times as much energy as the next lower number. Thus an earthquake of 7.5 on the Richter scale has a magnitude that is 10 times greater than that of a 6.5 earthquake.

Procedure

1. The data in Table 1 show the distances from four seismograph stations to epicenters of two of the aftershocks of the 1964 Alaskan earthquake, which occurred on March 27. Aftershocks are less severe quakes that follow the main quake.

Table 1

Distance between seismograph stations and epicenters of two March 29, 1964, aftershocks of the 1964 Alaskan earthquake.

Station	Distance to earthquake at 4:12 GMT* (km)	Distance to earthquake at 6:05 GMT* (km)
Bellingham, Wash.	1,938	2,280
Edmonton, Alberta, Canada	2,025	2,560
Pasadena, Calif.	3,440	3,600
Honolulu, Hawaii	4,420	3,880

*Twenty-four hour clock time is used: 4:12 GMT is 4:12 a.m. Greenwich Mean Time.

Using the data, a compass, and the directions below, mark the locations of the epicenters of these two earthquakes on the equidistant map (Figure 5) on Worksheet 1. This is a specially prepared map in which all distances away from Anchorage are correct. On most maps, the distances are distorted.

Figure 3 shows how to set a compass to have an accurate radius. Part of the distance is determined on the right-hand scale, and part on the left-hand scale, shown on the bottom of Worksheet 1.

From the four seismograph stations, draw circles or arcs having a radius equal to the distances to each of the earthquake epicenters. Although data are only needed from three stations to locate an epicenter, seismologists usually plot more than three distances for greater accuracy.

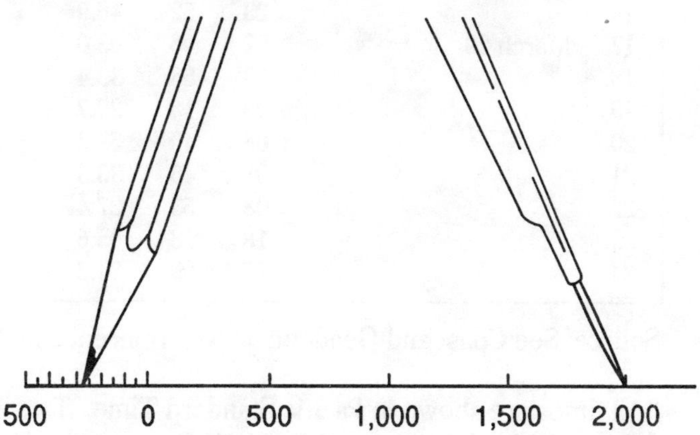

Figure 3. Compass is set at 2,280 km by measuring 2,000 km to the right of zero and measuring 280 km to the left.

2. Determine the latitude and longitude of the epicenters of these two aftershocks, and circle them in Table 2.

3. Using the latitudes and longitudes given in Table 2, plot on the map (Figure 6) on Worksheet 2 the location of the epicenter of the main shock (no. 1 on Table 2) and the 23 aftershocks (nos. 2-24) of the 1964 Alaskan earthquake. Number each epicenter. The locations of these 24 were determined by seismologists using the same method that you just used to determine the location of the two aftershock epicenters. Hundreds of aftershocks occurred after the main shock. Table 2 lists some of the stronger ones.

Table 2

Time, location, and magnitudes of the main shock and some of the aftershocks of the 1964 Alaskan earthquake.

	Date	Hr.	Min.	Sec.*	Latitude	Longitude	Magnitude (Richter scale)
1.	March 27	19	36	13.0	61.0°N	147.5°W	8.4
2.		22	43	56.4	58.3°N	151.3°W	6.1**
3.		23	10	21.4	58.8°N	149.6°W	6.2
4.	March 28	00	33	46.1	58.0°N	151.2°W	6.5
5.		01	01	00.4	56.5°N	152.0°W	6.2
6.		01	52	54.0	59.7°N	146.6°W	6.2
7.		03	08	26.5	60.1°N	148.5°W	6.2
8.		04	03	17.3	60.3°N	146.6°W	5.1
9.		04	20	48.8	56.5°N	154.1°W	6.5
10.		06	47	38.7	60.4°N	146.5°W	6.3
11.		06	49	15.0	60.4°N	147.1°W	6.5
12.		12	29	05.9	59.8°N	148.9°W	6.6
13.		17	09	36.4	59.9°N	149.2°W	5.2
14.		20	12	16.7	60.2°N	145.4°W	5.3**
15.		22	04	43.4	56.2°N	154.2°W	5.8
16.		23	52	45.9	56.2°N	154.1°W	4.9
17.	March 29	02	08	03.0	60.0°N	148.5°W	5.2**
18.		03	56	33.4	57.9°N	151.5°W	5.0**
19.		08	18	29.7	60.3°N	146.2°W	4.9
20.		08	40	59.3	59.8°N	146.9°W	5.8
21.		08	45	33.3	59.8°N	147.0°W	5.3**
22.		08	53	27.4	60.4°N	146.0°W	5.0**
23.		18	18	05.6	56.6°N	153.0°W	6.6
24.		23	09	34.2	59.8°N	145.9°W	6.2

Source: See Coast and Geodetic Survey reference in **Additional resources.**

*All times are shown in Pacific Standard Time. Twenty-four hour clock time is used: 19:36 is 7:36 p.m.
**Magnitudes by Coast and Geodetic Survey; all others are from Pasadena, California.

Questions

1. Where were the two aftershocks located? Give your answer in directions and kilometers from Anchorage.
2. What is the general distribution of the epicenters you plotted on Worksheet 2?
3. What was the sequence of these earthquakes in time? Did they occur progressively farther away from the epicenter of the main shock, like falling dominoes?
4. What were the magnitudes of the two aftershocks?
5. According to the theory of **plate tectonics,** the outer portion of the Earth is divided into large, moving sections or plates. Zones of earthquake activity coincide with plate boundaries. A **subduction zone** is a region where two plates converge or come together and one plate descends beneath the other. Subduction is occurring along the southern coast of Alaska where the Pacific plate is descending beneath Alaska on the North American plate, as shown in Figure 4. How does the theory of plate tectonics explain the origin of the 1964 Alaskan earthquake?

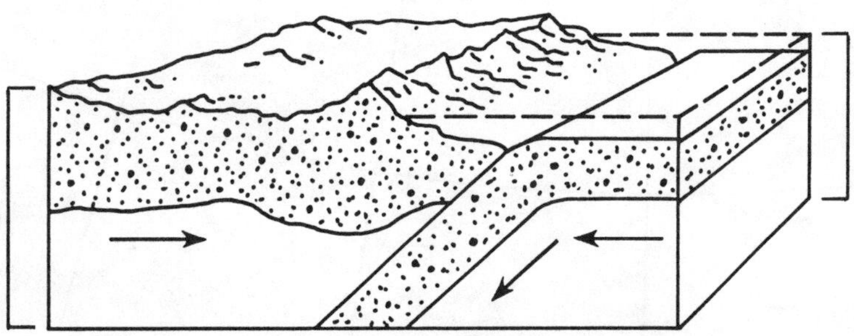

Figure 4. Schematic drawing showing the Pacific plate being subducted beneath the North American plate.

Additional resources

American Geological Institute. *Investigating the Earth.* 4th ed. Prepared by William H. Matthews et al. Boston: Houghton Mifflin Co., 1984.

Christman, Robert A. *Understanding and Interpreting the Bellingham Seismograms of the 1964 Alaskan Earthquake.* Bellingham, Wash.: Creative Dimensions, 1974. 40 p.

Coast and Geodetic Survey. *Preliminary Report, Prince William Sound Alaskan Earthquakes, March-April 1964.* Prepared by Seismology Division, U.S. Dept. of Commerce, Coast and Geodetic Survey, Washington, D.C. (April 1964).

Hanse, Wallace, et al. "The Alaskan Earthquake, March 27, 1964: Investigations and Reconstructions." *U.S. Geological Survey Professional Paper 541* (1966).

Hodgson, John. *Earthquakes and Earth Structure.* Englewood Cliffs, N.J.: Prentice Hall, 1964.

Epicenters of the 1964 Alaskan Earthquake

Worksheet 1

Name ....
Class ....
Date ....

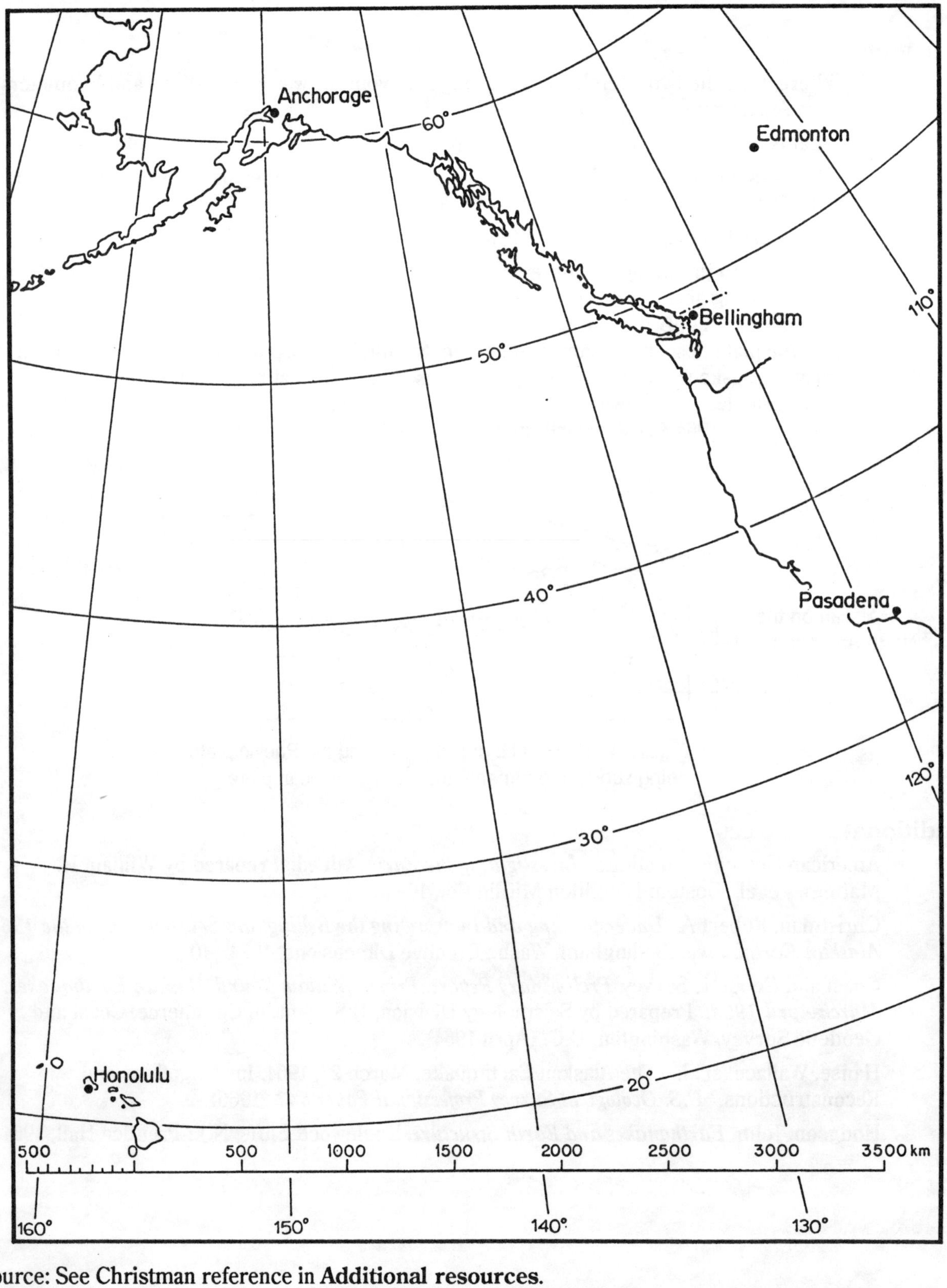

Source: See Christman reference in **Additional resources.**

Figure 5. Equidistant map centered on Anchorage to be used for locating epicenters of the Alaskan earthquakes. Use scale only for measuring distances to and from the vicinity of Anchorage.

Name ....

Class ....

Date ....

Worksheet 2

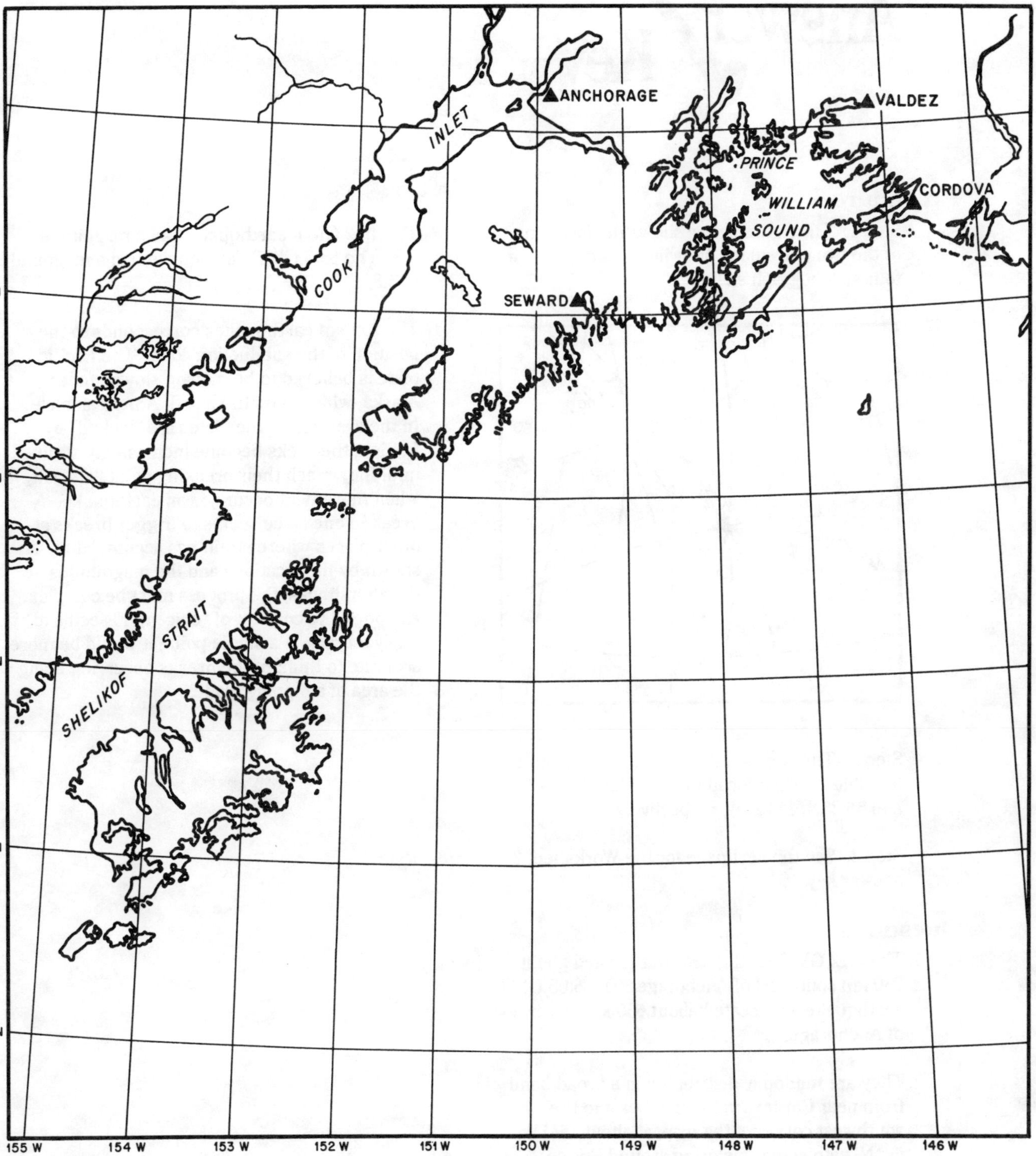

Source: See Christman reference in **Additional resources.**

Figure 6. Map of earthquake area.

Answer Key

Procedure

Step 1. The figure below shows the locations of the two epicenters, showing arcs drawn from four seismograph stations.

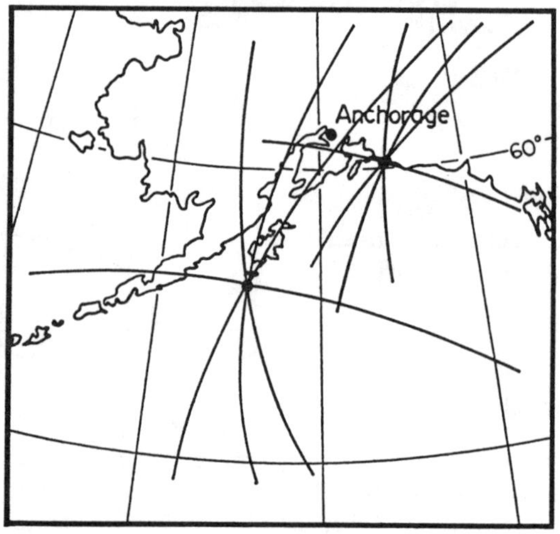

Step 2. The two aftershocks are nos. 14 and 15 in Table 2. Their locations are 60.2° N, 145.4° W and 56.2° N, 154.2° W, respectively.

Step 3. See the next page for the Worksheet 2 answer key.

Questions

1. The 4:12 GMT earthquake was located about 250 km southeast of Anchorage. The 6:05 GMT earthquake was located about 650 km southwest of Anchorage.

2. They are randomly distributed in a broad band from near Cordova in the northeast to the southwest corner of the map, at about 154° W, 56° N, with some clustering around Prince William Sound.

3. No sequence is evident. The order seems random, even though this does not seem logical.

4. The 4:12 GMT earthquake had a magnitude of 5.3. The 6:05 GMT earthquake had a magnitude of 5.8.

5. The zone of earthquakes corresponds to the position of the subduction zone. The Pacific plate is believed to be moving slowly under Alaska, which is on the North American plate. In the area where the plates are "locked" by friction, the rocks become increasingly strained until they reach their breaking point. The jolt, when movement occurs, is an earthquake. A break in one place seems to trigger breaks at other places where strain has accumulated. As shown by the locations and the magnitudes of the aftershocks, the process must be complex. Although the concept of "epicenter" seems to imply a break at a single point, it would be more accurate to think of the breaks occurring along the area of a dipping plane.

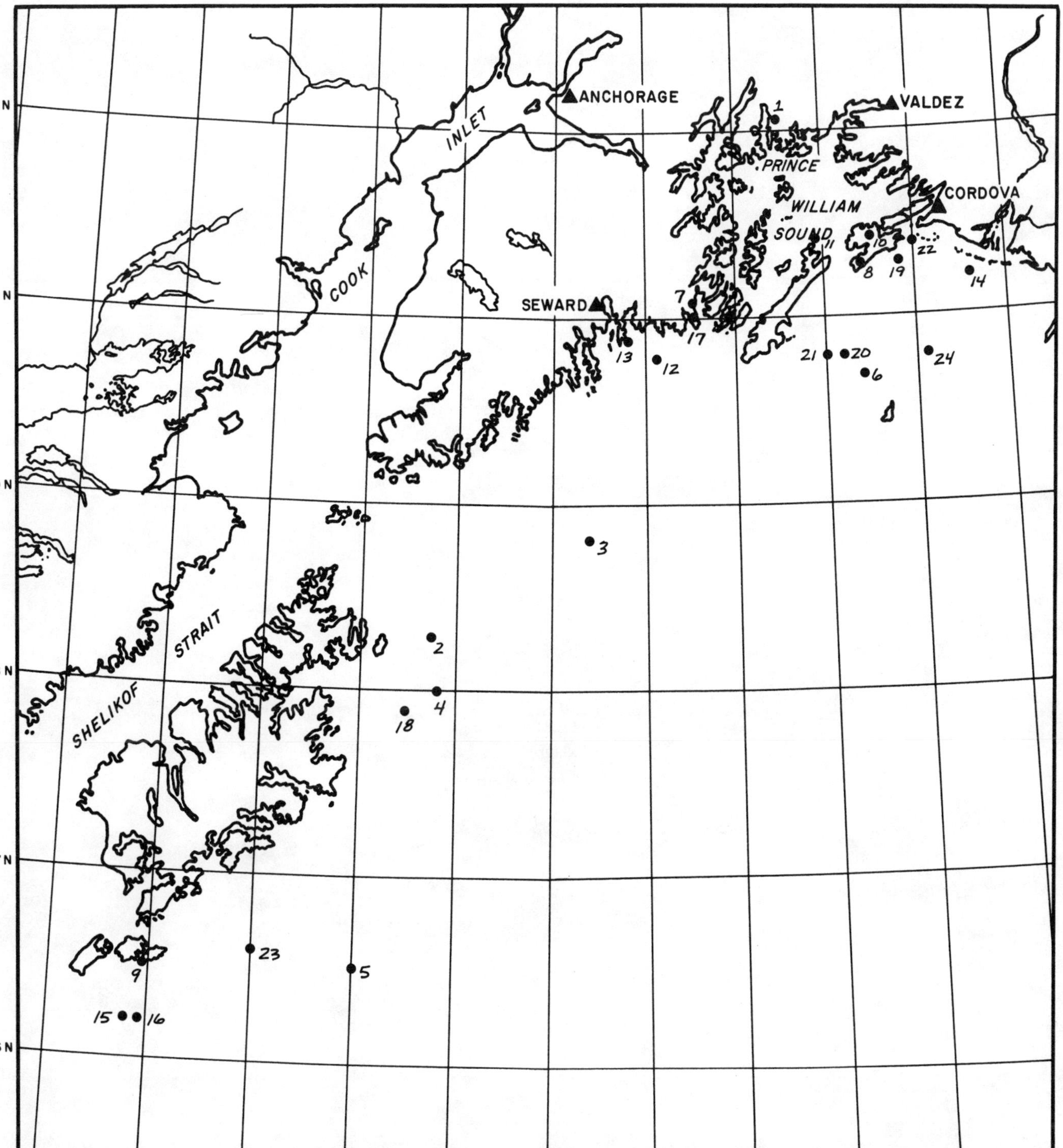

Source: See Christman reference in **Additional resources.**

Figure 6. Map of earthquake area.

Geophysical Investigations

THOMAS MCGUIRE

Briarcliff High School
Briarcliff Manor, New York

Objectives

To recognize four major landforms.

To plot gravitational data on a graph.

To draw isolines on a profile that shows subsurface temperatures.

To locate earthquake foci using coordinates of longitude and depth.

To see how landforms, gravitational measurements, and subsurface temperatures support the plate tectonic model of the Earth.

Time required

One class period

Materials

No additional materials are needed.

Background

Note: Students should know how to plot data on a graph and how to draw isolines.

By plotting data on the worksheet, you will see how three physical measurements — gravity, internal temperature, and earthquake depths — are related to surface features on the Earth. The data are hypothetical, however, the relationships that you will see are typical for the planet Earth. Both graphs on the worksheet represent one east-west traverse.

Procedure

Part I. Landforms

Label each of the following landforms shown on the topographic profile on Figure 1 on the worksheet.

mountains The highest elevations of the granitic crust.

ocean ridge A mostly submerged basaltic mountain range on the ocean floor.

ocean trench A deep trough on the ocean floor with the greatest ocean depths, often near a continent.

continental shelf The margins of the granitic continents covered by shallow ocean water.

Part II. Gravity

The strength (magnitude) of the Earth's pull (gravity) at any location on the Earth depends upon the density of the underlying bedrock. Within the Earth, relatively dense rock, such as basalt, causes the gravity to be relatively strong at the surface. Granitic rocks are less dense than basalt, so the gravitational pull at the surface is less than where the crust is composed of granitic rocks.

The data in Table 1 show local variations in gravity for the region shown on the worksheet. The magnitude of the gravity is presented as the difference between the local gravitational attraction and the expected value. The magnitude is expressed as a percentage of the expected value. A positive value indicates a stronger than expected gravitational attraction; a negative value indicates a weaker than expected gravitational attraction; and a value of zero indicates a local gravitational attraction equal to the expected gravitational attraction.

Table 1
Gravity data.

Longitude	Magnitude of gravity (%)	Longitude	Magnitude of gravity (%)
Prime meridian	+0.1	55°W	-0.2
5°W	+0.2	60°W	-0.3
10°W	+0.3	65°W	-0.4
15°W	+0.2	70°W	-0.1
20°W	+0.1	75°W	-0.2
25°W	+0.1	80°W	0.0
30°W	+0.1	85°W	+0.1
35°W	0.0	90°W	+0.1
40°W	0.0	95°W	+0.1
45°W	-0.1	100°W	+0.1
50°W	-0.1		

1. Plot the data in Table 1 on the gravity traverse section of the worksheet.
2. Draw a smooth line connecting the points.

Part III. Temperature

The Earth contains a great deal of heat. Some of it is left over from the formation of the Earth about 4.6 billion years ago, and some is from the ongoing decay of radioactive elements within the Earth. Some heat escapes to the surface by the direct contact of heated material with surrounding material, a process called **conduction.** Heat is also transferred by **convection,** the fluid movement of rock material caused by differences in density.

On the worksheet, the circled numbers below the topographic profile represent temperatures, expressed in degrees Celsius, at various locations inside the Earth. Use these numbers to draw **isotherms,** isolines connecting points of equal temperature, at intervals of 100° C.

Part IV. Earthquake depth

Earthquakes are most common in zones of crustal activity. According to the theory of **plate tectonics**, the outer portion of the Earth is divided into large, moving sections or plates. Zones of earthquake activity coincide with plate boundaries. Earthquakes that occur at great depth indicate areas where relatively cool material is being drawn into the Earth. These areas are known as **subduction zones**, long, narrow belts where plates descend into the Earth. Ocean trenches are areas where subduction occurs. Zones of shallow earthquakes indicate boundaries between spreading plates, such as ocean ridges, where hot basaltic material moves up to form new crust.

Using the data in Table 2, mark each earthquake **focus**, its point of origin, on the worksheet. Use an "X" to show each focus.

Table 2
Locations of earthquake foci.

Longitude	Depth (km)	Longitude	Depth (km)
10°W	4	11°W	3
75°W	17	65°W	36
78°W	8	12°W	7
44°W	3	78°W	11
8°W	6	85°W	9
69°W	28	71°W	22
58°W	0	5°W	8
72°W	12	70°W	23
96°W	8	66°W	2

Questions

1. According to your worksheet, at what geographic feature is gravity the strongest?
2. What igneous rock is the most common in the ocean crust?
3. Compared to granitic rocks, is basalt a low-density or a high-density rock?
4. What igneous rock is the most abundant in the continental crust?
5. Why is gravity slightly weaker on the continents than over the oceans?
6. What are isotherms?
7. Where does hot material seem to be rising to the surface?
8. Where is cool material being drawn into the Earth?
9. In what two places are earthquakes most common?
10. Where are earthquakes abundant but shallow?
11. Why are there more deep earthquakes near the ocean trenches than near other landforms?

Worksheet

Name ....

Class ....

Date ....

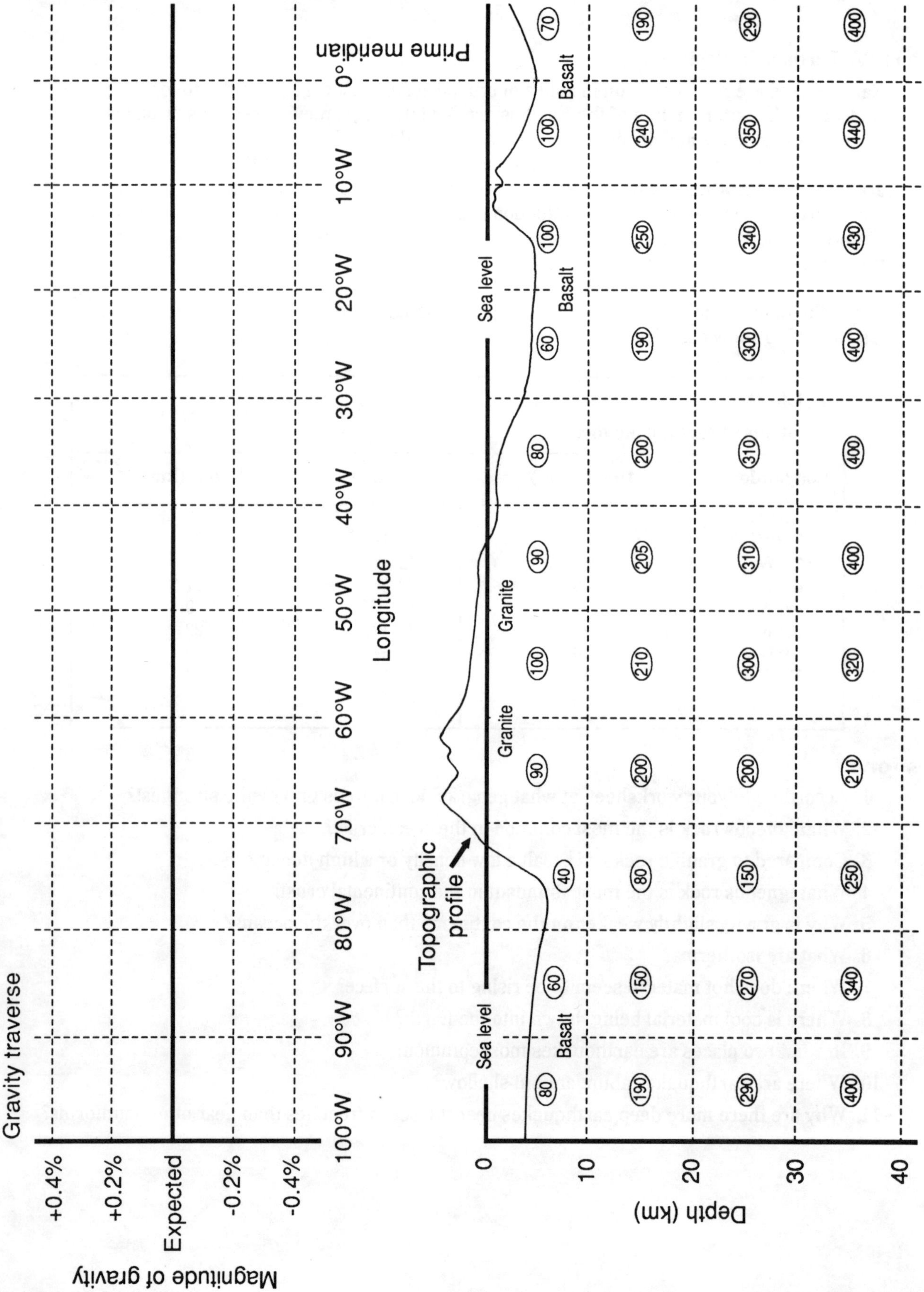

Figure 1. Geophysical data for a hypothetical east-west traverse. Circled numbers are temperatures (°C).

Answer Key

Procedure

See the next page for the worksheet answer key.

Questions

1. The strongest gravity readings occur over the ocean ridge.

2. Ocean crust is composed primarily of basalt.

3. When compared to granitic rocks, basalt is a high-density rock.

4. Continental crust is mainly composed of granitic material.

5. Continental crust, which is composed of granitic material, is less dense than oceanic crust, which is composed of basalt. Granitic rocks of the continent therefore exert a weaker gravitational attraction than oceanic crust exerts.

6. Isotherms are lines that connect points of equal temperature.

7. The isotherms show that hot material seems to be rising to the surface at the ocean ridge.

8. The isotherms show that cool material is being drawn down into the Earth at the ocean trench.

9. Earthquakes are most common along ocean trenches and ocean ridges.

10. Shallow earthquakes are abundant around ocean ridges.

11. Ocean trenches are areas where one plate is being drawn down (subducted) beneath another. This movement causes cool rock to be pushed down into the Earth. Rock must be cool to fracture. Only in subduction zones is rock at great depth cool enough to fracture and cause deep earthquakes.

Answer Key

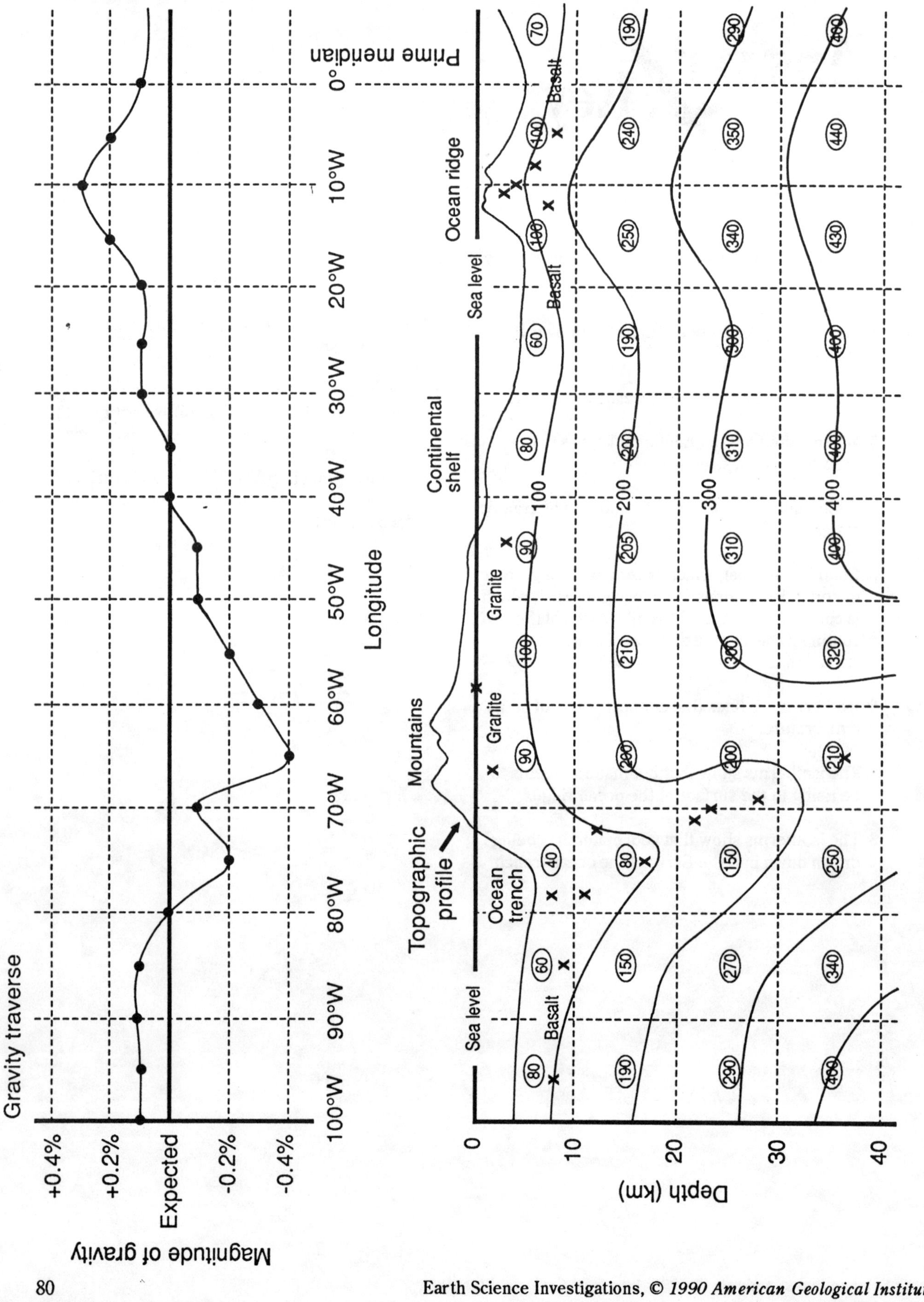

Figure 1. Geophysical data for a hypothetical east-west traverse. Circled numbers are temperatures (°C).

How Did the North Atlantic Form?

PAUL JOHNSTON
Emporia State University
Emporia, Kansas

Objective

To visualize how the North Atlantic Ocean basin formed and to understand how different ages of bedrock reveal its history.

Time required

One class period

Materials

per student
2 pencils or markers, different colors: one color should be blue.
Scissors
Blue paper for map base

Background

The Atlantic Ocean, as we know it, did not exist 200 million years ago. At that time, the present-day continents of Africa, the Americas, Europe, and Asia were parts of a large single supercontinent. Geologists think that this supercontinent, called Pangea, existed from approximately 300 to 200 million years ago.

Our understanding of the breakup of Pangea and the formation of ocean basins is based on the theory of **plate tectonics**. According to this theory, the outer portion of the Earth is composed of large irregular sections or plates. At their boundaries, plates may move apart (zones of divergence), together (zones of convergence), or past one another (transform faults). Plates grow or decrease in size as a result of movement and interaction with other plates. Ocean basins gradually develop in zones of divergence by repeated episodes of **rifting**, or fracturing, and volcanic activity. Magma, or molten rock, moves up through the rifts and cools to form new rock material that enlarges the ocean floor along plate margins. This process is known as **sea-floor spreading**, and the line along which the spreading occurs is the **axis of spreading**. As sea-floor spreading causes new rock material to be added along the spreading plate margins, the continents riding on the plates gradually move farther apart.

At present, sea-floor spreading occurs along a chain of undersea ridges that extends throughout the world's ocean basins. Continued sea-floor spreading along the mid-Atlantic ridge causes the Atlantic Ocean basin to grow a few centimeters a year.

Figure 1 is a map showing the relative locations of the continents before the rifting that formed the North Atlantic began. The dashed lines indicate rift zones, or fracture zones, along which the sea floor spread and the continents moved. The dotted areas indicate the Paleozoic orogenic belt, a mountain chain of folded and faulted rocks that are believed to have resulted from plate collisions.

In this investigation, you will cut the map along the dashed lines to help you visualize the progressive stages of the formation of the North Atlantic. Use the geologic time scale (Figure 2) to see when the stages occurred.

Procedure

Part I. Jurassic time (208 to 144 million years ago)

1. On the map (Figure 1), outline the continents in one color, and shade the water blue.
2. Cut along the dashed line from 1 through 2 to 4 and to the right through 5. Fit the pieces back together on top of blue paper.
3. Slide the top half of the map to the left until A and A′ touch. Notice that the North American continent is still attached to Europe. These continents slipped as one unit along a transform fault between Africa and Europe (dashed line between 4 and 5). Sea-floor spreading occurred along the segment between 2 and 4.

Part II. Early Cretaceous time (144 to 98 million years ago)

1. Cut the short dashed line along the west edge of Spain.
2. Cut through the top of the paper from 7 to 6. Leave Spain attached to Europe at 6.
3. During this time, North America began to slide northwest relative to Greenland. Slide North America toward the upper left until B and B′ touch. It moved along the transform fault that follows the general direction of the Labrador Sea. Notice that this movement opens the Atlantic between the Grand Banks and Spain.

Part III. Late Cretaceous time (98 to 66 million years ago)

1. Separate North America from Greenland by about one inch, thus opening the Labrador Sea. During the Late Cretaceous, as North America began to pull away from Greenland, the Labrador Sea transform fault became an axis of sea-floor spreading.
2. Across the Atlantic, the Bay of Biscay, which is in line with the Labrador Sea, also began to open. Holding the paper flat, lift the tab that has Spain on it and move it toward Africa. This movement causes a wrinkle to develop in the paper at 6. With your thumb, press the wrinkle so that it stays in place. This wrinkle is the Pyrenees Mountains. As Spain moved south relative to Europe, the Bay of Biscay opened.

Part IV. Tertiary time (66 to 2 million years ago)

1. Greenland remained joined with Europe until early in the Tertiary. The spreading stopped in the Labrador Sea and began in the Norwegian Sea on the other side of Greenland. Separate Greenland from Europe by cutting along the dashed line through 8.

2. Keeping North America and Greenland about one inch apart, move them as one unit to the left, separating Greenland from Europe.

Part V. Quaternary time (2 million years ago to the present)

The axis of spreading, which is widening the North Atlantic, follows the line from 8 along the Norwegian Sea and southward between North America and Africa.

The dashed line between 2 and 3 represents the spreading axis of the South Atlantic between South America and Africa. This segment began to separate early in the Cretaceous. It, along with the North Atlantic, is still widening.

Questions

1. During what geologic time period did the bedrock of the ocean floor just off the west coast of Spain form?
2. The oldest bedrock on the Atlantic Ocean floor would be found adjacent to the coasts of which two continents?
3. The rocks forming the floor of the Bay of Biscay are the same age as those on the floor of what other body of water?
4. Today, does Greenland move along with North America as part of the North American plate or with Europe as part of the European plate?
5. The Appalachian Mountains of the eastern U.S. extend northeast to the island of Newfoundland where they appear to end abruptly. Where would you look to find the northeastern continuation of this mountain system?
6. The Paleozoic orogenic belt was formed at the time that Pangea existed. The present-day spreading axis between North America and Africa does not exactly follow the boundary between the plates whose movements resulted in the formation of this belt. During the Paleozoic, the southeastern part of the United States would have been part of which present-day continent?
7. The map that you cut to show how the present-day land masses formed from the splitting up of an earlier larger continent did not have Iceland on it. Why was Iceland omitted?
8. Could the Pyrenees Mountains be considered part of the Paleozoic orogenic belt? Explain.

Additional resources

Carpenter, John R., and Astwood, Philip M. *Plate Tectonics for Introductory Geology.* Dubuque: Kendall/Hunt Publishing, 1983.

Wilson, J. Tuzo, ed. *Continents Adrift.* Readings from *Scientific American.* San Francisco: W.H. Freeman and Co., 1972.

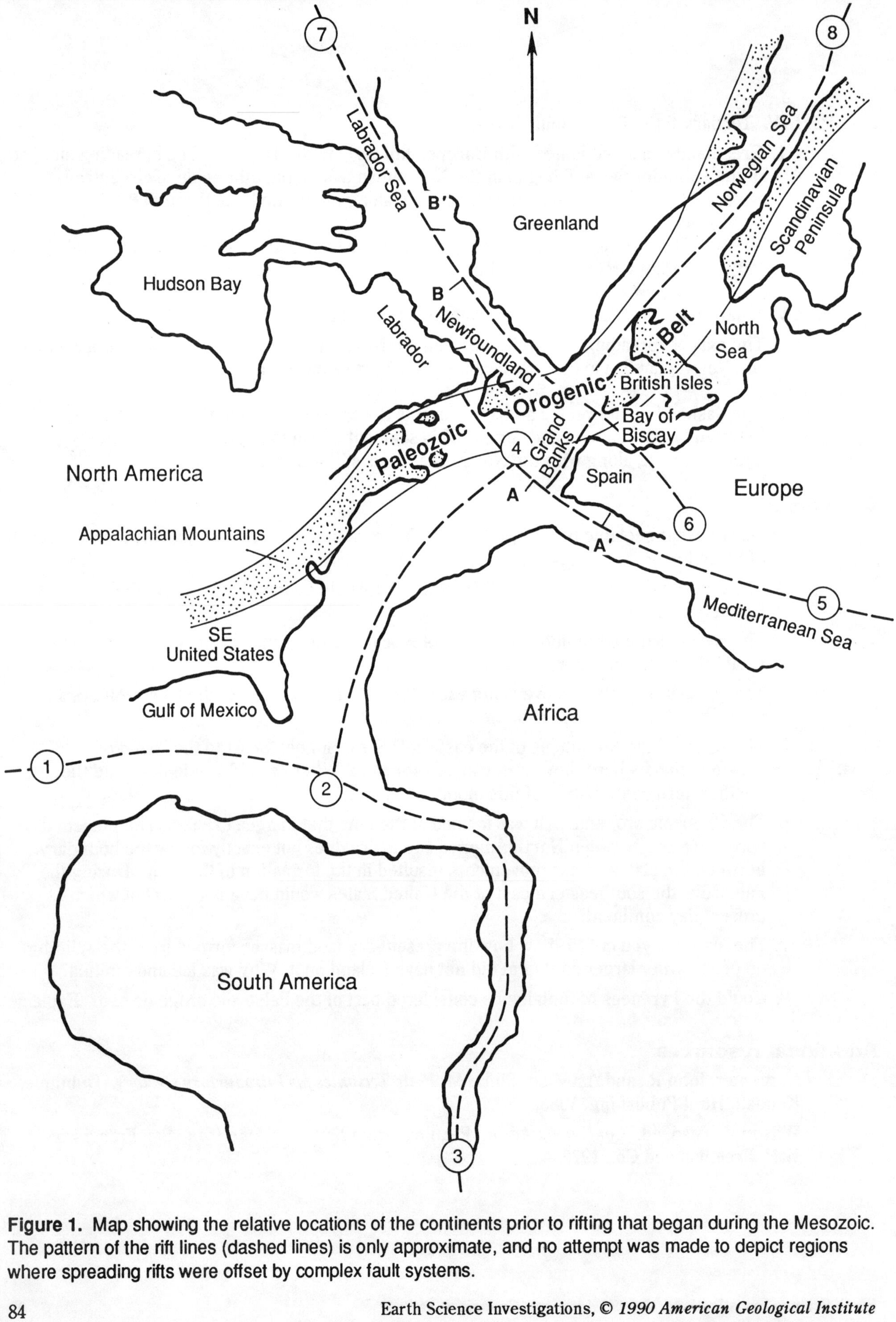

Figure 1. Map showing the relative locations of the continents prior to rifting that began during the Mesozoic. The pattern of the rift lines (dashed lines) is only approximate, and no attempt was made to depict regions where spreading rifts were offset by complex fault systems.

Era	Period	Age estimates (millions of years ago)
Cenozoic	Quaternary	2
	Tertiary	66
Mesozoic	Cretaceous	144
	Jurassic	208
	Triassic	245
Paleozoic	Permian	286
	Pennsylvanian	320
	Mississippian	360
	Devonian	408
	Silurian	438
	Ordovician	505
	Cambrian	570
Precambrian		

Figure 2. Geologic time scale. The space shown for each period corresponds to its relative duration, except for the Precambrian. Notice the short duration of the current period, the Quaternary. The time scale is open at the bottom, because the Precambrian represents 90% of geologic time. To show the relative duration of this vast span of time would require eight additional pages.

Answer Key

Questions

1. The rocks just off the west coast of Spain formed during the Early Cretaceous. At that time, when Spain broke away from the Grand Banks, sea-floor basalt would have begun to flow into the space between them.

2. Since the North Atlantic began to open by North America splitting away from Africa, you would expect to find the oldest ocean floor rock off the coasts of these continents. The age of the oldest rock is Jurassic.

3. The Bay of Biscay began to open at the same time as the transform fault in the Labrador Sea became a spreading axis. The rocks forming the floor of the Bay of Biscay are thus the same age as those forming the floor of the Labrador Sea. The age of the rocks in both is Late Cretaceous.

4. During the Early Cretaceous, Greenland remained fastened to Europe as North America separated from Europe. By early in the Tertiary, spreading had stopped in the Labrador Sea. When spreading began in the Norwegian Sea, Greenland started to move away from Europe, along with North America, as part of the North American plate.

5. The Appalachian Mountains are part of the Paleozoic orogenic belt. This same folded mountain system continues on the other side of the modern Atlantic as the mountains of the northern British Isles and the Scandinavian Peninsula.

6. That portion of the United States that lies to the southeast of the Appalachians would have been part of the African continent during the Paleozoic. The North American plate did not separate from the African plate until the Jurassic.

7. Iceland is not a fragment of Pangea. It is volcanic in origin and formed from lava extruded in the spreading axis between Greenland and Europe.

8. No, the Pyrenees formed at a later time, during the Late Cretaceous. They are not part of the Paleozoic orogenic belt.

Investigating Tides

DARREL HOFF
Harvard-Smithsonian Center for Astrophysics
Cambridge, Massachusetts

Objective

To investigate the pattern of tides and to construct a mental model to explain them, using the phases of the moon.

Time required

One class period

Materials

No additional materials are needed.

Background

The familiar ebb and flow of tides often dictates the cycle of life for people who make their living from the sea. Clam diggers depend on low tides. Some ports are usable only during high tides. Devastating storms often occur at high tides. Knowledge of tidal patterns helped in planning the Normandy invasion in France in 1944 and at Inchon during the Korean War.

Tides have also played a part in the development of life. The cycle of tides has left its imprint on the growth layers of the fossil shells of tidal-basin animals, providing clues to early Earth-moon history.

Gravitational attraction among the Earth, moon, and sun causes tides. The height of tides is governed by the positions of these planetary bodies in relation to each other as the Earth-moon system revolves around the sun.

It takes about 29.5 days for the moon to make one revolution around the Earth. During this orbit, the side of the moon facing the sun is illuminated. The angle from which the illuminated side of the moon is seen from the Earth changes, causing the moon to appear as different shapes called phases. Figure 1 shows the four main phases of the moon: new moon, first quarter, full moon, third quarter.

At full and new moons, the sun, moon, and Earth are lined up, and the combined gravitational pull of the moon and sun results in extreme high and low tides. Such tides are called **spring tides**; they seem to "spring" up higher during these phases.

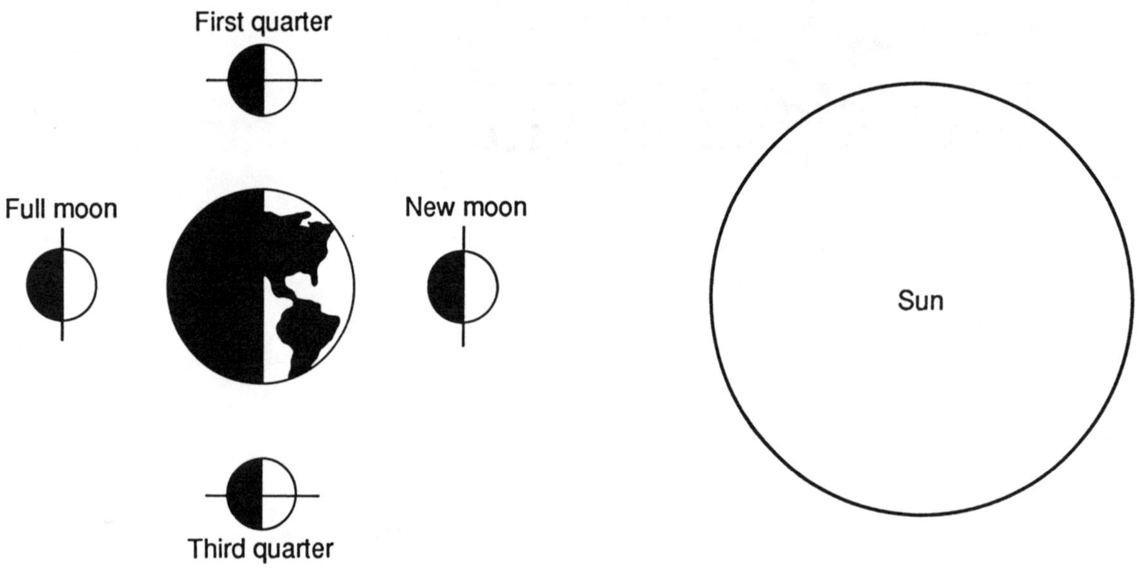

Figure 1. Phases of the moon (not to scale).

At first and third quarter moons, the sun and moon are at right angles to each other relative to the Earth, and each one's gravitational pull on the Earth is in opposition to the other's. The result is high tides that are the lowest and low tides that are the highest during the lunar cycle. Such tides are called **neap tides**.

Procedure

Students will use Tables 1 and 2 to complete Table 3 on the worksheet and answer the questions.

Questions

1. Examine Table 1, which shows the tides for Pointe de Grave, France, for three months in 1976. The first column lists the date; the second column, the time of the high and low tide, given in 24-hour clock time (for example, 0121 is 1:21 a.m., 1354 is 1:54 p.m.); and the third column is the height of the tide (in feet) above a mean reference point. How many tides do you typically find each day?
2. Do you always find that many each day?
3. How long is it from one high tide to the next low tide?
4. How long is it from a high tide to the next high tide?
5. How long is it from the first high tide of one day to the first high tide of the next day?
6. Scan down the table and see if you can determine any pattern in how tidal heights change during a month. For example, as the high tides get lower from day to day, what happens to the height of the low tides?
7. For the month of January, find the two consecutive dates when the first highest high tides occur. Now find the two consecutive dates when the next highest high tides occur. How many days is it between these tides?
8. When the high tides are the highest, how would you describe the height of the low tides? What is this type of tide called?
9. Find the date halfway between the two sets of spring tides. Describe the heights of the high and low tides on the date. What is this type of tide called?

Table 1

Tide tables for Pointe de Grave, France, 1976.

January

Day	Time	Height (ft.)	Day	Time	Height (ft.)
1	0437	17.1	16	0420	16.2
TH	1040	3.6	F	1022	3.9
	1658	16.7		1640	16.2
	2302	3.8		2244	3.9
2	0519	17.4	17	0501	17.1
F	1124	3.4	SA	1104	3.4
	1740	16.9		1721	16.9
	2345	3.6		2325	3.3
3	0600	17.6	18	0541	17.9
SA	1205	3.4	SU	1146	3.0
	1820	16.7		1801	17.4
4	0025	3.8	19	0007	3.0
SU	0639	17.2	M	0622	18.2
	1245	3.6		1227	2.8
	1858	16.4		1843	17.6
5	0103	3.9	20	0049	3.0
M	0717	16.9	TU	0704	18.2
	1321	3.9		1311	2.8
	1935	16.1		1925	17.4
6	0138	4.4	21	0131	3.1
TU	0754	16.2	W	0747	17.9
	1356	4.4		1353	3.1
	2013	15.6		2011	16.9
7	0215	4.9	22	0214	3.4
W	0833	15.7	TH	0836	17.1
	1433	5.1		1436	3.6
	2056	15.1		2103	16.2
8	0251	5.6	23	0300	4.1
TH	0920	15.1	F	0932	16.2
	1513	5.7		1526	4.4
	2143	14.4		2202	15.4
9	0334	6.2	24	0354	5.1
F	1010	14.4	SA	1034	15.4
	1559	6.4		1625	5.4
	2238	13.8		2310	14.8
10	0427	6.7	25	0458	5.7
SA	1110	13.8	SU	1149	14.6
	1656	6.9		1736	5.9
	2345	13.5			
11	0529	7.1	26	0030	14.4
SU	1220	13.6	M	0616	6.1
	1804	7.1		1309	14.4
				1856	6.1
12	0056	13.6	27	0146	14.6
M	0640	6.9	TU	0737	5.7
	1329	13.6		1420	14.6
	1915	6.9		2013	5.7
13	0200	14.1	28	0251	15.1
TU	0749	6.4	W	0847	5.1
	1427	14.3		1520	15.3
	2019	6.2		2117	5.1
14	0254	14.8	29	0345	15.7
W	0847	5.7	TH	0944	4.4
	1518	14.9		1608	15.7
	2113	5.4		2209	4.4
15	0340	15.6	30	0430	16.4
TH	0937	4.9	F	1031	3.9
	1601	15.6		1650	16.2
	2200	4.6		2253	3.9
			31	0509	16.9
			SA	1112	3.4
				1727	16.6
				2331	3.6

February

Day	Time	Height (ft.)	Day	Time	Height (ft.)
1	0546	17.2	16	0522	18.4
SU	1149	3.3	M	1127	2.5
	1803	16.7		1743	18.2
				2348	2.5
2	0006	3.4	17	0604	19.0
M	0620	17.2	TU	1209	2.3
	1223	3.3		1824	18.5
	1836	16.7			
3	0041	3.6	18	0030	2.3
TU	0653	17.1	W	0645	19.2
	1257	3.4		1251	2.3
	1909	16.4		1906	18.4
4	0113	3.8	19	0113	2.5
W	0725	16.6	TH	0727	18.5
	1327	3.9		1333	2.6
	1940	16.1		1949	17.6
5	0143	4.3	20	0154	3.0
TH	0758	16.1	F	0812	17.4
	1358	4.4		1415	3.3
	2014	15.6		2037	16.6
6	0215	4.9	21	0238	3.8
F	0832	15.4	SA	0905	16.2
	1431	5.1		1502	4.3
	2053	14.9		2134	15.6
7	0249	5.6	22	0329	4.9
SA	0914	14.8	SU	1006	15.1
	1508	5.9		1559	5.6
	2139	14.3		2241	14.6
8	0330	6.2	23	0432	5.9
SU	1005	13.9	M	1121	14.1
	1555	6.7		1711	6.4
	2235	13.6			
9	0425	6.9	24	0006	13.9
M	1110	13.3	TU	0554	6.4
	1658	7.2		1251	13.6
	2350	13.1		1840	6.6
10	0537	7.2	25	0132	13.9
TU	1233	13.1	W	0723	6.2
	1819	7.2		1410	13.9
				2003	6.1
11	0112	13.5	26	0243	14.6
W	0701	6.9	TH	0837	5.4
	1350	13.5		1510	14.6
	1940	6.7		2107	5.2
12	0222	14.1	27	0335	15.4
TH	0814	5.9	F	0933	4.8
	1451	14.4		1557	15.4
	2045	5.6		2156	4.6
13	0317	15.3	28	0417	16.1
F	0912	4.9	SA	1017	4.1
	1540	15.4		1635	15.9
	2138	4.6		2235	3.9
14	0401	16.2	29	0452	16.6
SA	1002	3.8	SU	1054	3.6
	1622	16.2		1708	16.4
	2223	3.4		2311	3.6
15	0442	17.2			
SU	1045	3.0			
	1703	17.2			
	2306	2.8			

March

Day	Time	Height (ft.)	Day	Time	Height (ft.)
1	0525	16.9	16	0500	18.7
M	1127	3.3	TU	1103	2.3
	1741	16.7		1720	18.7
	2343	3.4		2324	2.3
2	0556	17.1	17	0540	19.4
TU	1158	3.3	W	1146	2.1
	1811	16.7		1802	19.0
3	0013	3.4	18	0007	2.1
W	0626	17.1	TH	0623	19.4
	1228	3.3		1228	2.1
	1840	16.6		1844	18.7
4	0044	3.6	19	0051	2.3
TH	0655	16.7	F	0706	18.5
	1258	3.6		1313	2.6
	1910	16.2		1928	17.9
5	0113	3.9	20	0134	2.8
F	0725	16.2	SA	0750	17.4
	1327	4.1		1355	3.3
	1940	15.9		2015	16.6
6	0142	4.4	21	0218	3.8
SA	0757	15.6	SU	0841	16.1
	1357	4.8		1442	4.4
	2013	15.3		2111	15.6
7	0214	5.1	22	0309	4.9
SU	0832	14.9	M	0943	14.8
	1432	5.4		1538	5.6
	2055	14.6		2218	14.4
8	0251	5.7	23	0412	5.9
M	0919	14.1	TU	1059	13.8
	1513	6.2		1650	6.6
	2147	13.8		2344	13.8
9	0339	6.6	24	0534	6.6
TU	1018	13.3	W	1231	13.3
	1611	7.1		1819	6.7
	2257	13.3			
10	0449	7.1	25	0113	13.8
W	1142	13.0	TH	0702	6.4
	1732	7.2		1352	13.6
				1941	6.2
11	0030	13.3	26	0223	14.4
TH	0617	6.9	F	0814	5.6
	1312	13.3		1450	14.4
	1901	6.7		2043	5.4
12	0151	13.9	27	0314	15.1
F	0740	5.9	SA	0909	4.9
	1422	14.1		1535	15.1
	2013	5.6		2131	4.8
13	0250	15.1	28	0354	15.7
SA	0843	4.8	SU	0952	4.3
	1515	15.3		1611	15.7
	2110	4.4		2210	4.1
14	0337	16.2	29	0427	16.2
SU	0935	3.6	M	1027	3.8
	1558	16.4		1643	16.2
	2158	3.3		2244	3.8
15	0418	17.6	30	0459	16.6
M	1020	2.8	TU	1059	3.4
	1639	17.7		1713	16.6
	2242	2.6		2315	3.4
			31	0528	16.7
			W	1129	3.4
				1743	16.7
				2345	3.4

Source: *Tide Tables - Europe and West Coast of Africa - 1976.* (Rockville, Md.: U.S. Dept. of Commerce, National Oceanic and Atmospheric Administration), 40.

10. Fill in the first column in Table 3 on the worksheet, listing the consecutive dates for the spring tides and the neap tides for January, February, and March.
11. Examine Table 2, which lists the phases of the moon for the same period covered in the tidal information in Table 1. In Table 3 in the second column, record the moon phase that corresponds to the tide dates you have listed.
12. What conclusions can you draw about what phases of the moon typically occur near the dates of the spring tides?
13. What phases of the moon occur near the dates of the neap tides?
14. What is the basic pattern of the tides?
15. (Optional) The full and the new moons are on our meridian at midnight and noon, though we cannot always see them at those times. One would expect that the spring high tides would occur at noon and midnight. From Table 3, choose a date when the moon is full.

    On the date you chose, what are the times for the high tides?

    Are these times near noon or midnight?

    Why don't the highest tides during spring tides occur when the moon is on the meridian? Note that the highest tide on January 1, a new moon date, actually occurs about seven hours before noon.

Table 2
Phases of the moon, January-March, 1976.

Date		Hour	Phase
January	1	15	New
	9	13	First quarter
	17	05	Full
	23	23	Third quarter
	31	06	New
February	8	10	First quarter
	15	17	Full
	22	09	Third quarter
	29	23	New
March	9	05	First quarter
	16	03	Full
	22	19	Third quarter
	30	17	New

Source: *Observer's Handbook - 1976.* (Toronto: Royal Astronomical Society of Canada). Published annually.

Additional resources

Abell, George O., et al. *Exploration of the Universe.* 5th ed. New York: W.B. Saunders Co., 1987.

Hicks, S.D. "Ocean Tides." *The Science Teacher* 32 (December 1965):11-14.

Runcorn, S.K. "Corals as Paleontological Clocks." *Scientific American* 215, no. 4 (October 1966): 26-33.

Name ....

Class ....

Date ....

Worksheet

Table 3
Relationships between moon phases and Earth tides.

	Date	Moon phase
Spring tides	....	....
	....	....
	....	....
	....	....
	....	....
	....	....
Neap tides	....	....
	....	....
	....	....
	....	....
	....	....
	....	....

Answer ✓ Key

Questions

1. Typically, four tides occur each day, two high and two low tides.

2. No, on some days there are only three tides.

3. The interval between one high tide and the next low tide is a little more than six hours.

4. The interval between one high tide and the next high tide is a little more than 12 hours.

5. The interval between the first high tide of one day and the first high tide of the next day is a little more than 24 hours — nearly 25 hours in some cases.

6. As the high tides get lower from day to day, the low tides get higher from day to day, and then the pattern reverses.

7. The first highest high tides occur on January 2 and 3. The next highest high tides occur on January 19 and 20. There are 15 days between these two tides.

8. When the high tides are the highest, the low tides are the lowest. This type of tide is called a spring tide.

9. January 10 and 11 have the lowest high tides. When the high tides are the lowest, the low tides are the highest. This type of tide is called a neap tide.

10. and 11. (Table 3)

Spring tides:

Date	Moon phase
Jan. 2-3	Near new
Jan. 19-20	Near full
Feb. 1-2	Near new
Feb. 17-18	Near full
Mar. 2-3	Near new
Mar. 17-18	Near full

Neap tides:

Date	Moon phase
Jan. 10-11	Near first quarter
Jan. 25-26	Near third quarter
Feb. 9-10	Near first quarter
Feb. 23-24	Near third quarter
Mar. 9-10	Near first quarter
Mar. 24-25	Near third quarter

12. A full or new moon occurs near the dates of the spring tides.

13. A first or third quarter moon occurs near the dates of the neap tides.

14. High tides are the highest and low tides are the lowest — spring tides — near the time of the full and new phases of the moon. High tides are the lowest and low tides are the highest — neap tides — near the time of the first and third quarter phases of the moon.

15. (Optional) There is a lag in the tidal system. The moon is not exactly at the new or full phase when the spring tides occur. In addition, the time when a new or full moon is on the meridian does not correspond to the time of the highest tide. For example, there was a full moon on January 17, and therefore a spring tide occurred near that time (January 19-20). Yet on January 17, the high tides occurred at 0501 and at 1721, not at midnight and noon. The reason for this time difference is that the tides slow the Earth's rotation.

    The Earth's tidal bulge points generally in the moon's direction, but this bulge is drawn ahead of the Earth-moon line because of the Earth's rotation. Frictional forces between the water and the surface of the Earth (particularly in shallow seas) are set up because the Earth rotates once in 24 hours, but the moon revolves in the same direction around the Earth in a synodic period of 29.5 days.

The consequences are that the high tides are pulled ahead of the Earth-moon line, resulting in a gradual slowing of the Earth's rotation. This slowing, though gradual, accumulates such that a constant running clock started 20 centuries ago compared with a clock geared to the rotation of the Earth, would have "gained" about three and a half hours during two thousand years. Careful examination of historical eclipse records supports this contention. Today we have accurate quartz time-keeping systems. Nearly every year, the National Bureau of Standards inserts a "leap" second at the end of June or at the end of December to keep the Earth (rotation) clock in time with precise time-keeping devices.

Was Mount St. Helens Moonstruck?

ROBERT A. CHRISTMAN
Western Washington University
Bellingham, Washington

Objective

To determine whether the straining of rocks from the gravitational forces of the moon and sun influenced the eruptions of Mount St. Helens.

Time required

One class period

Materials

No additional materials are needed.

Background

During the 10 months of 1980 when Mount St. Helens erupted sporadically, the Portland, Oregon, and Vancouver, Washington, newspapers reported that a relationship might exist between the volcanic eruptions and the phases of the moon. Various people predicted that Mount St. Helens would erupt during a particular time period, because the moon was full or new.

The actual moment when a volcano will erupt is difficult to predict. The May 18, 1980, eruption of Mount St. Helens followed an earthquake and a major landslide of the summit area on the volcano. If the earthquake had not occurred, this eruption may have been delayed for several days. As pressures develop and increase within a magma chamber, a volcano may be on the verge of eruption over a period of time. Thus, it is nearly impossible to determine when an eruption will occur and what additional stress (or release of the stress) will cause an eruption.

Scientists have considered whether tidal forces may be related to the timing of an eruption. When the moon and sun are lined up with the Earth, or the moon is closest to the Earth, or both, the gravitational pull causes very high and very low tides on the Earth. This alignment occurs during two phases of the moon: **full moon,** when the Earth is between the sun and the moon, and **new moon,** when the moon is between the Earth and the sun. The point in the moon's orbit when it is closest to the Earth is called its **perigee**. Because these conditions cause extreme tides, it has been argued that these same forces might cause rocks within or overlying a volcano to fracture or break. Thus, gravitational forces might have triggered eruptions of Mount St. Helens on the days when the highest and lowest tides occurred.

To test this hypothesis, you will use data in Tables 1 and 2. Table 1 lists the dates of eruptions of Mount St. Helens. Table 2 on the worksheet lists the dates and magnitudes of all the minus tides that occurred at the town of Aberdeen, Washington, between March 14 and December 31, 1980. A minus tide occurs when a low tide is below "mean low water," the average of all low tides. Aberdeen is located on the Pacific Ocean, about 145 km (90 miles) northwest of Mount St. Helens. It is the closest location for which complete tidal records for 1980 have been published by the National Oceanic and Atmospheric Administration.

Procedure

1. Make a prediction before you examine the facts. Do you think that the eruptions of a volcano can be predicted by the phases of the moon or the variations in its distance from the Earth or both?
2. On Table 2 on the worksheet, draw a square around the dates when the low tides were -2.0 and lower. Circle the dates when the low tides were between -1.0 and -2.0. The dates you have marked represent the times when the gravitational pull must have been the greatest.
3. Examine Table 1. Compare the dates when eruptions occurred (Table 1) with your marks when minus tides of -2.0 and lower occurred (Table 2). On Table 2, mark each eruption with a date and an arrow. Look for correlations.

Table 1

Important eruptions of Mount St. Helens in 1980.

Month	Day	Eruption
March	27	First eruption (small)
April	1	First large eruption (minor)
April	8	Minor eruptions for five hours
April	13	Many minor eruptions
May	18	Catastrophic eruption (major)
May	25	Major eruption
June	12-13	Major eruption
July	22	Major eruption
August	7	Major eruption
October	16-18	Major eruption

Questions

1. What is your prediction?
2. How many sequences of minus tides (-1.0 and lower) occurred?
3. How many of the major eruptions occurred when the moon's pull was strong?
4. How many times (sequences) was the moon's pull strong but eruptions did not occur?
5. Do you see a relationship between sequences of days with very low tides and times when eruptions occurred?
6. What is the correlation between eruptions and minus tides?
7. Do you think the moon caused Mount St. Helens to erupt?
8. What are the problems in predicting volcanic eruptions?

Additional resources

Dzurisin, Daniel. "Influence of Fortnightly Earth Tides at Kilauea Volcano, Hawaii." *Geophysical Research Letters* 7 (1980): 925-928.

Lipman, P.W., and Mullineaux, D.R. "The 1980 Eruptions of Mount St. Helens." *U.S. Geological Survey Professional Paper 1250* (1981).

Name ....

Class ....

Date ....

Worksheet

Table 2

Minus tides recorded at Aberdeen, Washington, between March 14 and December 31, 1980.

Date*	Magnitude of lowest tide (ft.)	Moon data
March		
14	-0.4	
15	-0.7	New moon, Perigee
16	-0.7	
20	-0.4	
April		
14	-0.5	Perigee
15	-1.3	New moon
16	-1.6	
17	-1.7	
18	-1.3	
19	-0.8	
20	-0.1	
29	-0.1	
30	-0.4	Full moon
May		
1	-0.6	
2	-0.6	
3	-0.6	
4	-0.4	
5	-0.2	
12	-0.5	Perigee
13	-1.4	
14	-2.1	New moon
15	-2.3	
16	-2.2	
17	-1.8	
18	-1.2	
19	-0.5	
27	-0.1	
28	-0.7	
29	-1.1	Full moon
30	-1.3	
31	-1.4	
June		
1	-1.4	
2	-1.2	
3	-0.9	
4	-0.5	
5	-0.1	
9	-0.1	Perigee
10	-1.0	
11	-1.8	
12	-2.2	New moon
13	-2.3	
14	-2.2	
15	-1.8	
16	-1.3	
17	-0.6	
25	-0.2	
26	-0.8	
27	-1.3	
28	-1.7	Full moon
29	-1.8	
30	-1.8	
July		
1	-1.6	
2	-1.2	
3	-0.7	
8	-0.3	
9	-1.0	
10	-1.5	
11	-1.8	
12	-1.9	New moon
13	-1.8	
14	-1.5	
15	-1.0	
16	-0.5	
25	-0.7	
26	-1.3	
27	-1.7	Full moon
28	-2.0	
29	-1.9	
30	-1.6	Perigee
31	-1.1	
August		
1	-0.3	
6	-0.1	
7	-0.6	
8	-0.9	
9	-1.2	
10	-1.3	New moon
11	-1.3	
12	-0.9	
13	-0.5	
23	-0.4	
24	-1.0	
25	-1.4	
26	-1.6	Full moon
27	-1.5	Perigee
28	-1.0	
29	-0.4	
30	-0.1	
September		
6	-0.1	
7	-0.3	
8	-0.4	
9	-0.3	New moon
22	-0.3	
23	-0.6	
24	-0.7	Full moon
25	-0.8	Perigee
26	-1.0	
27	-1.0	
28	-0.6	
29	-0.1	
October		
22	-0.1	
23	-1.0	Full moon, Perigee
24	-1.5	
25	-1.6	
26	-1.3	
27	-0.8	
28	-0.1	
November		
20	-0.4	
21	-1.2	Perigee
22	-1.6	Full moon
23	-1.6	
24	-1.2	
25	-0.6	
December		
8	-0.1	
19	-0.3	Perigee
20	-0.9	Full moon
21	-1.2	
22	-1.2	
23	-0.9	
24	-0.3	

* For dates not listed in this table, the lowest tides were not minus tides. They were zero feet or higher.

Answer ✓ Key

Procedure

Depending upon the students' backgrounds, some discussion on the origin of tides and the relationship of the moon to the Earth may be needed.

Students may ask why the correlation between the moon's positions and the recorded tides is not better. Barometric pressure and the presence of strong winds blowing offshore or onshore may modify the magnitude of minus tides. Those factors would reduce the correlation.

Steps 2 and 3. See the next page for the worksheet answer key.

Questions

1. Answers will vary. This is a guess.

2. Thirteen sequences of minus tides (-1.0 and lower) occurred.

3. Two major eruptions occurred. Of the six major eruptions, a correlation seems to exist for two of them — May 18 and June 12-13. No correlation between the other eruptions and minus tides is apparent.

4. In 11 sequences of minus tides (-1.0 and lower), when the moon's pull was strong, eruptions did not occur.

5. A relationship may exist. Of the two sequences when the -2.0 and lower minus tides lasted for three consecutive days, an eruption took place both times. The major eruption of May 18 followed the May 14-16 minus tides days. The major eruption of June 12-13 occurred during a corresponding three-day period of -2.2 and lower tides.

6. The correlation is poor. The moon's position cannot be used accurately to predict an eruption. What would people have thought if the scientists had predicted a major eruption on July 28 when the tide was -2.0? No eruption occurred.

7. Students will probably say no. The possibility of a relationship, however, between the moon's pull, which causes a stress on the rocks of the Earth, and the timing of an eruption cannot be ignored completely. The relationship might be the "straw-that-broke-the-camel's-back" type. If the conditions within a volcano are such that an eruption is about to occur, the effect of the moon's pull might be the triggering factor. No one knows for sure. Dzurisin (see **Additional resources** for reference) has found a close correlation with tides for some volcanoes.

8. The data are inconclusive. Probably no single cause for a volcanic eruption exists. It is likely that different factors, for which scientists have no information, act together to cause an eruption.

Answer Key

Table 2

Minus tides recorded at Aberdeen, Washington, between March 14 and December 31, 1980.

Date*	Magnitude of lowest tide (ft.)	Moon data
March		
14	-0.4	
15	-0.7	New moon, Perigee
16	-0.7	
20	-0.4	
April		
14	-0.5	Perigee
15	-1.3	New moon
16	-1.6	
17	-1.7	
18	-1.3	
19	-0.8	
20	-0.1	
29	-0.1	
30	-0.4	Full moon
May		
1	-0.6	
2	-0.6	
3	-0.6	
4	-0.4	
5	-0.2	
12	-0.5	Perigee
13	-1.4	
14	-2.1	New moon
15	-2.3	
16	-2.2	
17	-1.8	
18	-1.2	
19	-0.5	
27	-0.1	
28	-0.7	
29	-1.1	Full moon
30	-1.3	
31	-1.4	
June		
1	-1.4	
2	-1.2	
3	-0.9	
4	-0.5	
5	-0.1	
9	-0.1	Perigee
10	-1.0	
11	-1.8	
12	-2.2	New moon
13	-2.3	
14	-2.2	
15	-1.8	
16	-1.3	
17	-0.6	
25	-0.2	
26	-0.8	
27	-1.3	
28	-1.7	Full moon
29	-1.8	
30	-1.8	
July		
1	-1.6	
2	-1.2	
3	-0.7	
8	-0.3	
9	-1.0	
10	-1.5	
11	-1.8	
12	-1.9	New moon
13	-1.8	
14	-1.5	
15	-1.0	
16	-0.5	
25	-0.7	
26	-1.3	
27	-1.7	Full moon
28	-2.0	
29	-1.9	
30	-1.6	Perigee
31	-1.1	
August		
1	-0.3	
6	-0.1	
7	-0.6	
8	-0.9	
9	-1.2	
10	-1.3	New moon
11	-1.3	
12	-0.9	
13	-0.5	
23	-0.4	
24	-1.0	
25	-1.4	
26	-1.6	Full moon
27	-1.5	Perigee
28	-1.0	
29	-0.4	
30	-0.1	
September		
6	-0.1	
7	-0.3	
8	-0.4	
9	-0.3	New moon
22	-0.3	
23	-0.6	
24	-0.7	Full moon
25	-0.8	Perigee
26	-1.0	
27	-1.0	
28	-0.6	
29	-0.1	
October		
22	-0.1	
23	-1.0	Full moon, Perigee
24	-1.5	
25	-1.6	
26	-1.3	
27	-0.8	
28	-0.1	
November		
20	-0.4	
21	-1.2	Perigee
22	-1.6	Full moon
23	-1.6	
24	-1.2	
25	-0.6	
December		
8	-0.1	
19	-0.3	Perigee
20	-0.9	Full moon
21	-1.2	
22	-1.2	
23	-0.9	
24	-0.3	

*For dates not listed in this table, the lowest tides were not minus tides. They were zero feet or higher.

Layers of the Atmosphere

JACK FEARING
Lincoln Junior High School
Hibbing, Minnesota

Objective

To discover how the atmosphere can be divided into layers based on temperature changes at different heights, by making a graph.

Time required

Two class periods

Materials

No additional materials are needed.

Background

Note: Students should know how to plot data on a graph.

The atmosphere can be divided into four layers based on temperature variations. The layer closest to the Earth is called the **troposphere**. Above this layer is the **stratosphere**, followed by the **mesosphere**, then the **thermosphere**. The upper boundaries between these layers are known as the **tropopause**, the **stratopause**, and the **mesopause**, respectively.

Temperature variations in the four layers are due to the way solar energy is absorbed as it moves downward through the atmosphere. The Earth's surface is the primary absorber of solar energy. Some of this energy is reradiated by the Earth as heat, which warms the overlying troposphere. The global average temperature in the troposphere rapidly decreases with altitude until the tropopause, the boundary between the troposphere and the stratosphere.

The temperature begins to increase with altitude in the stratosphere. This warming is caused by a form of oxygen called ozone (O_3) absorbing ultraviolet radiation from the sun. Ozone protects us from most of the sun's ultraviolet radiation, which can cause cancer, genetic mutations, and sunburn. Scientists are concerned that human activity is contributing to a decrease in stratospheric ozone. Nitric oxide, which is in the exhaust of high-flying jets, and chlorofluorocarbons (CFCs), which are used as refrigerants, may contribute to ozone depletion.

At the stratopause, the temperature stops increasing with altitude. The overlying mesosphere does not absorb solar radiation, so the temperature decreases with altitude. At the mesopause, the temperature begins to increase with altitude, and this trend continues in the thermosphere. Because the atmosphere is so thin, a thermometer cannot measure the temperature accurately, and special instruments are needed.

Procedure

1. Table 1 contains average temperature readings at various altitudes in the Earth's atmosphere. Plot this data on the graph (Figure 1) on the worksheet, and connect adjacent points with solid lines. This profile provides a general picture of temperature conditions in the atmosphere; at any given time and place, however, the actual temperature may deviate from the average values, particularly in the lower atmosphere.
2. Label the different layers of the atmosphere and the separating boundaries between each layer.
3. Mark the general location of the ozone layer.

Table 1
Average temperature readings at various altitudes.

Altitude (km)	Temperature (°C)	Altitude (km)	Temperature (°C)
0	15	52	-2
5	-18	55	-7
10	-49	60	-17
12	-56	65	-33
20	-56	70	-54
25	-51	75	-65
30	-46	80	-79
35	-37	84	-86
40	-22	92	-86
45	-8	95	-81
48	-2	100	-72

Questions

1. What is the basis for dividing the atmosphere into four layers?
2. Does the temperature increase or decrease with altitude in the troposphere? Stratosphere? Mesosphere? Thermosphere?
3. What is the approximate height and temperature of the tropopause, stratopause, and mesopause?
4. What causes the temperature to increase with height through the stratosphere and decrease with height through the mesosphere?
5. What causes the temperature to decrease with height in the troposphere?

Additional resources

Stanger, Robert L., and Perkins, Otho E. *Cambridge Work-A-Text in Earth Science.* New York: Globe Book Co., 1981, p. 133-135.

Namowitz, Samuel N., and Stone, Donald B. *Earth Science: The World We Live In.* 4th ed. New York: Van Nostrand Reinhold, American Book Co., 1969, p. 496-498.

Lutgens, Frederick K., and Tarbuck, Edward J. *The Atmosphere: An Introduction to Meteorology.* 4th ed. Englewood Cliffs, N.J.: Prentice Hall, 1989. 491 p.

Name ....

Class ....

Date ....

Worksheet

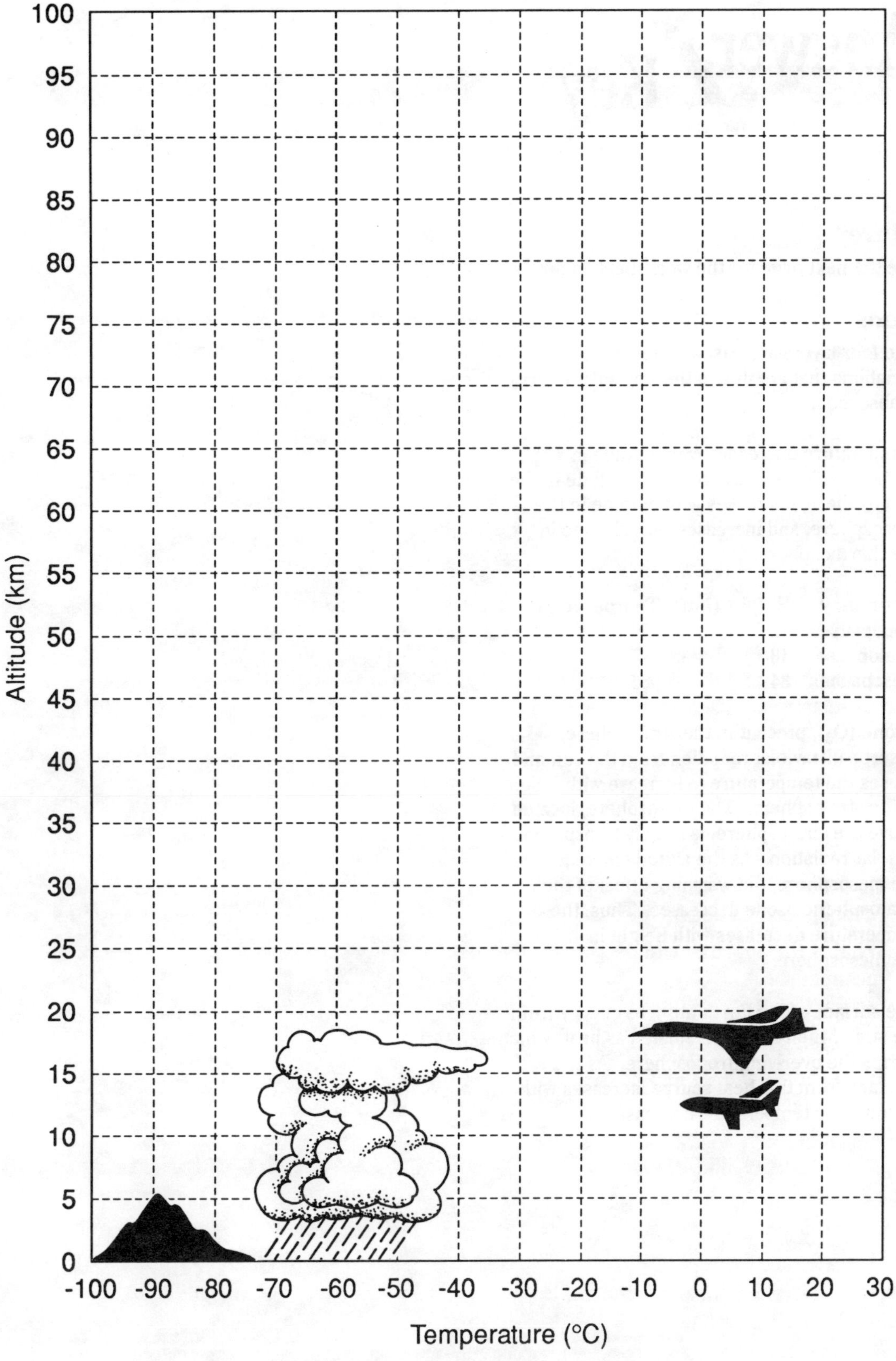

Figure 1. Graph of temperature at various altitudes.

Answer Key

Procedure

See the next page for the worksheet answer key.

Questions

1. The four layers are based on temperature variations that are due to the way solar energy is absorbed.

2. Temperature decreases with altitude in the troposphere, increases with altitude in the stratosphere, decreases with altitude in the mesosphere, and increases with altitude in the thermosphere.

3.

Boundary	Height (km)	Temperature (°C)
tropopause	12	-56
stratopause	48	-2
mesopause	84	-86

4. Ozone (O_3), present in the stratosphere, absorbs ultraviolet radiation from the sun and causes the temperature to increase with height in the stratosphere. The mesosphere, located above the stratosphere, is largely transparent to solar radiation. As the altitude increases in the mesosphere, the warming effect of the stratospheric ozone decreases. Thus, the temperature decreases with height in the mesosphere.

5. The surface of the Earth absorbs energy from the sun. Much of it is reradiated as heat, which warms the overlying troposphere. As the distance from this heat source increases with altitude, the temperature decreases.

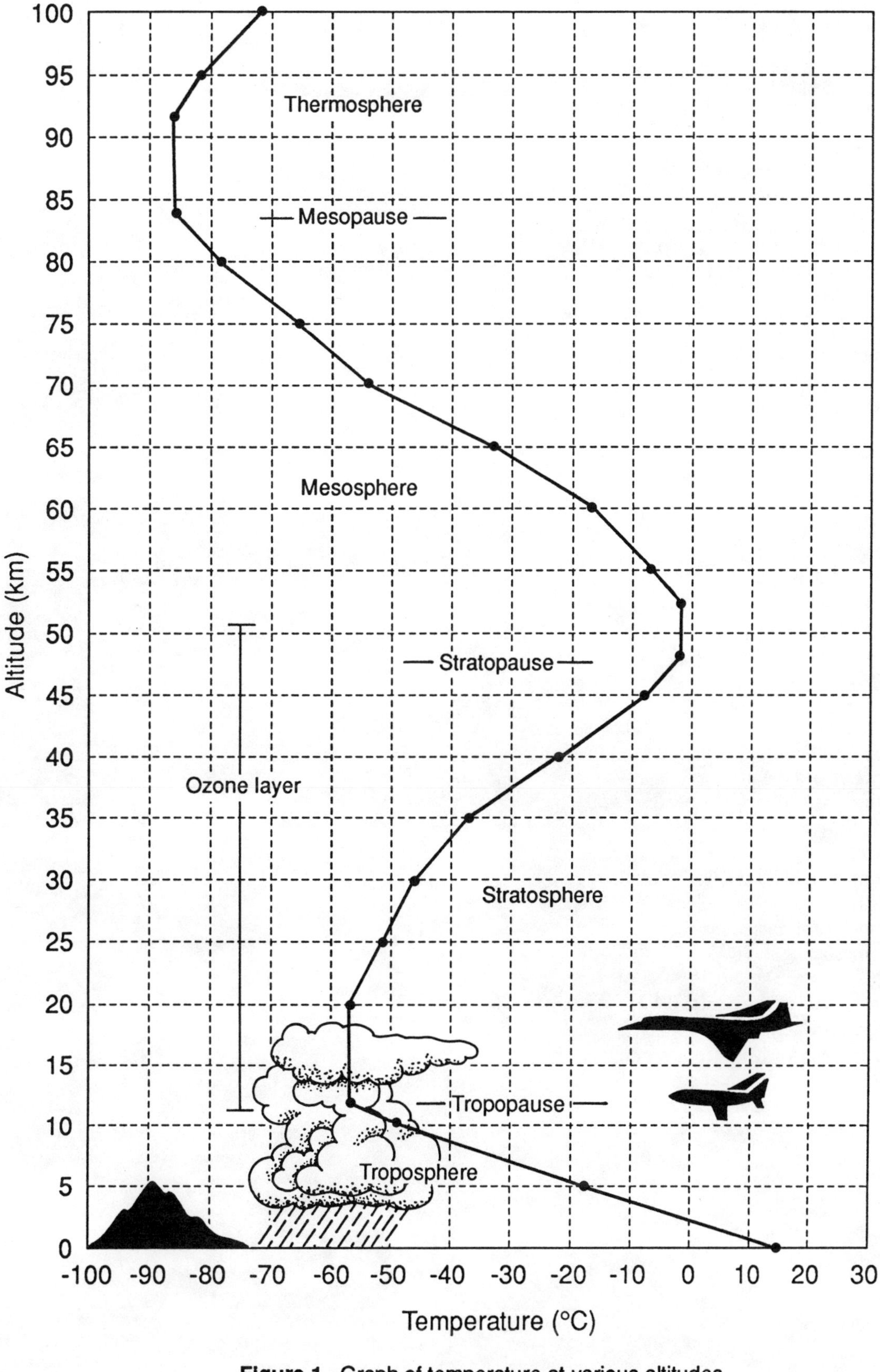

Figure 1. Graph of temperature at various altitudes.

Micro-Weather Patterns

JEANNE A. ENDRIKAT & DONALD S. EHRENBERGER

Lake Braddock Secondary School
Burke, Virginia

Objective

To determine whether various environments can affect wind speed, temperature, and relative humidity patterns at different heights above the ground and at different times of the day.

Time required

Two class periods: class 1 will be outside gathering data, and class 2 will be inside analyzing them.

Materials

per group of 2-4 students
Meter stick
2 thermometers
Masking tape
Beaufort wind scale or a wind meter
Sling psychrometer or a wet-and-dry bulb thermometer board
Graph paper

Background

Note: Students should know how to draw bar and line graphs.

Various surfaces and environments may influence temperature and relative humidity readings. Dark- and light-colored surfaces radiate heat differently. Different environments such as wooded sites, open sites, a north-facing slope, or a south-facing one, as well as the time of day and weather, will influence results. Over the years, changes in these environments may determine which plant and animal species can survive in them.

Procedure

As a class, select several environments near the school for determining temperature and relative humidity readings. Some suggested settings might be a gravel area, concrete sidewalk, black top surface, open grassy area, woods, metal surface, exposed soil or sand, and north-facing slope.

Part I. Gathering data (class 1)

1. Each student group should choose a different environment to investigate.

2. Calibrate the thermometers and the sling psychrometers so that all thermometers read the same temperature. To calibrate the thermometers, compare each to a room average temperature. Add or subtract the number of degrees below or above the room average temperature to each of the readings for that specific thermometer. You may wish to use a permanent marking pen to mark a change on the backing of the thermometer. **CAUTION:** ***Sling psychrometers break easily. Be careful when you handle them.***

3. Tape one thermometer to the meter stick so that the bottom of the bulb will be exactly one cm from the end. Tape the second thermometer to the same side of the meter stick so that the bottom of the bulb will be exactly 99 cm from the end. The bulb end will be up (Figure 1).

4. Outside at each location, allow one minute for the thermometers to adjust to the temperatures; be sure the thermometers are not in direct sunlight. Turn the meter stick if necessary. Take the temperature at one cm and at 99 cm. Record the group's observations of an environment in Table 1 on the worksheet.

5. Take readings for wind speed and relative humidity, and record them in Table 1. Fill in the remaining data.

Figure 1. Diagram showing two thermometers attached to the meter stick.

Part II. Analyzing data (class 2)

1. After reviewing the class data, graph some of the results and discuss the patterns and trends that were observed. Some suggested bar graphs might be wind speed vs. environment, temperature vs. environment, and relative humidity vs. environment.

2. If there are other classes that gathered the same data that day, discuss and graph their results, too. Suggested line graphs might be wind speed vs. time, one-cm temperature vs. time, 99-cm temperature vs. time, and relative humidity vs. time. Each environment could be represented by a different color line.

Questions

1. Does the type of environment appear to influence the relative humidity? Explain.
2. How does the type of environment affect the one-cm temperature? The 99-cm temperature? Describe any patterns you see.
3. Are the two temperatures of an environment the same? What seems to be the relationship between height and temperature?
4. What effect does wind speed have on your results?
5. What relationship do you think may exist between sky conditions and your results?

6. How does the time of day affect temperature and relative humidity?
7. If there are results from other classes, describe any changes you see happening throughout the day. Look at the highest and lowest temperatures, relative humidities, wind speeds, and sky conditions.
8. If a warm front had passed through your area that day, how would you expect it to influence your results?

Additional resources

Namowitz, Samuel N., and Spaulding, Nancy E. *Earth Science.* 9th ed. Lexington: D.C. Heath and Co., 1989.

Ramsey, William L., et al. *Modern Earth Science.* New York: Holt, Rinehart and Winston, 1989.

Worksheet

Name ....

Class ....

Date ....

Table 1
Conditions of various environments.

Environment	Time	Sky condition	Wind speed (mph)	Temperatures (°C) 1 cm	99 cm	Relative humidity

Answer Key

Questions

1. Probably yes, a wooded area would be expected to have a higher relative humidity than an open, sunny area.

2. Darker surfaces should have higher temperatures than lighter surfaces. Open surfaces should be warmer than wooded or shaded surfaces.

3. No, the height trends will probably vary from class to class as well as with general weather conditions.

4. A calm day should allow temperatures to warm up more and the relative humidity to be higher than a windy day.

5. A cloudy day will reduce the temperatures.

6. Earlier in the day, the temperatures should be lower, and the relative humidity should be higher.

7. Again, earlier in the day, the temperatures should be lower, and the relative humidity should be higher. Wind speed may increase throughout the day.

8. Before a warm front, the temperature will be lower. Temperatures will increase as the front passes through the area.

Modeling Crystal Growth in Magmas

Susan Reynolds Ballinger
Wauwatosa, Wisconsin

Objective

To observe and compare crystal growth and crystal size in a model that simulates magma that cools slowly and magma that cools quickly.

Time required

Fifteen minutes for each of three consecutive days
Forty-eight hours are needed for complete evaporation of the saltwater solution. A standing floor fan helps speed evaporation.

Materials

per student
Saltwater solution: 19 ml in a beaker
Graduated cylinder
Glass petri dish with lid: if possible, petri dish bottom should not have divisions in it.
2 labels for petri dish
Hand lens
Metric ruler

Background

Igneous rocks form from **magma**, or molten rock. Magma becomes solid rock by cooling. The rate of cooling determines crystal size of the component minerals. **Intrusive rocks** are formed from magma that crystallizes deep below the Earth's surface. The magma cools slowly and develops large crystals. **Extrusive rocks**, such as lava, are formed from magma that reaches the Earth's surface. The magma cools quickly and develops small or no crystals.

In this investigation, you will make a crystal-forming model of magma. Saltwater solution represents magma. The cooling of magma is represented by the evaporation of water, leaving salt crystals behind. Rocks are modeled by the collection of salt crystals. Two rates of cooling are represented by using two different initial amounts of saltwater solution.

Procedure

The teacher should make a saturated saltwater solution by dissolving two parts table salt (NaCl) into one part hot tap water, stirring until no more salt will go into solution. To simulate hot magma, the solution should be kept in a hot-water bath on a hot plate until it is distributed to the students. **CAUTION:** ***Some hot plates may get hot enough to cause burns without showing any signs of being hot.***

1. Label a petri dish lid "Magma outside the Earth." Label a petri dish bottom "Magma inside the Earth." Put your name on both labels.

2. Pour 4 ml of saltwater solution into the "Magma outside the Earth" petri dish. Pour 15 ml of saltwater solution into the "Magma inside the Earth" petri dish.

3. Observe the contents of each dish and record your observations beside "Day 0" in Table 1 on the worksheet.

4. Place uncovered petri dishes in a place where they will not be disturbed for 24 hours.

5. Observe your petri dishes after 24 hours. Record any changes, such as liquid or solid state, color, or crystal pattern, in Table 1 beside "Day 1."

6. Repeat step 4.

7. Repeat step 5, recording your observations in Table 1 beside "Day 2."

8. Use your hand lens to observe the sizes of crystals in both dishes.

9. While looking through the hand lens, position a metric ruler so you can read the markings through the petri dish.

10. Choose the six largest individual crystals in each dish. Measure the widest part of these crystals. Record the measurements in Table 2 on the worksheet.

11. Calculate the average diameter for both sets of crystals (add the six measurements in each column and divide the total by six). Record both averages.

Questions

1. In this model, what represents liquid rock? Solid rock?
2. How many days did it take for the "Magma outside the Earth" to become completely solid?
3. How many days did it take for the "Magma inside the Earth" to become completely solid?
4. When a magma becomes solid rock, cooling takes place. Your model represents cooling by evaporation. Which solution represents faster cooling magma? Slower cooling magma?
5. Which petri dish produced the crystals with the largest average diameter?
6. In general, as the length of crystallization time increases, what happens to the average crystal size?
7. How could you identify an igneous rock that cooled inside the Earth?
8. What will you look for in an igneous rock that cooled quickly?
9. Why does magma inside the Earth cool slower than magma outside the Earth?
10. In each dish, did your crystals grow separately or join to other crystals?
11. What pattern of mineral crystals would you expect to see in rocks?
12. Which crystals represent intrusive rocks? Extrusive rocks?

Name ....
Class ....
Date ....

Worksheet

Table 1
Record of crystal observations.

	Magma outside the Earth	Magma inside the Earth
Day 0		
Day 1		
Day 2		

Table 2
Widest dimension of six crystals in millimeters.

Crystal	Magma outside the Earth	Magma inside the Earth
1		
2		
3		
4		
5		
6		
Average		

Answer ✓ Key

Questions

1. The saltwater solution represents liquid rock. Salt crystals represent solid rock.

2. "Magma outside the Earth" became solid in one day.

3. "Magma inside the Earth" became solid in two days.

4. "Magma outside the Earth" was the faster cooling magma. "Magma inside the Earth" was the slower cooling magma.

5. "Magma inside the Earth" had the largest crystals.

6. The crystals become larger.

7. An igneous rock that forms inside the Earth will cool slowly and develop large crystals.

8. An igneous rock that cools quickly will not have large crystals. It might not have any crystals at all.

9. Magma on the Earth's surface is cooled by contact with land, air, or water. Magma inside the Earth is like a hot drink inside a thermos. The heat cannot escape as rapidly, so the magma cools slowly.

10. In "Magma outside the Earth," most crystals were touching other crystals. In "Magma inside the Earth," several of the large crystals were growing together in a clump.

11. Crystals would be interlocking, touching each other.

12. "Magma inside the Earth" (larger crystals) represents intrusive rocks. "Magma outside the Earth" (smaller crystals) represents extrusive rocks.

Rocks! Rocks! Rocks!

DONALD L. BIRDD
Buffalo State College
Buffalo, New York

Objective

To demonstrate and observe the rock cycle through a series of simulation activities that show the processes of mechanical weathering, erosion, and sedimentation, and the formation of sedimentary, metamorphic, and igneous rocks.

Time required

Four class periods

Materials

per student
Safety goggles
Laboratory apron
Pocket pencil sharpener
4-6 crayons of the same color. Red, green, blue, and yellow are recommended for each student group.
Wax paper: 30 cm x 30 cm (12 in. x 12 in.)
Newspaper to cover a flat work surface
Envelope

per group of 4-6 students
Hot plate
2 pairs of tongs
3 aluminum foil squares, heavy duty: 45 cm x 45 cm (18 in. x 18 in.)
2 aluminum foil disposable pie trays: 10-cm (4-in.) diameter, or a similar-sized rectangular tray
Glass petri dish or finger bowl

per class
3 vises (minimum number), or 2 6-in. or larger C-clamps per group
Wooden boards, 2 per vise or per 2 C-clamps: 2.5 cm x 12.5 cm x 20 cm (1 in. x 5 in. x 8 in.)

Background

Rocks and rock formations occur at or near the Earth's surface. However, they may be covered by buildings, roads, soil, and water. Look over the rock specimens available in your classroom. Talk about the rocks that you see where you live. Discuss the types of rocks that you see in road cuts.

Think about these questions: How did these rocks form? How long did it take? Where did they come from? Are more rocks forming at the present time? What causes rocks to break into small pieces? What about layers... and streaks? Write down other questions you may have about rocks and their formation. You can answer many of these questions by learning about the **rock cycle**, events involving the formation, alteration, destruction, and reformation of rocks as a result of natural processes (Figure 1).

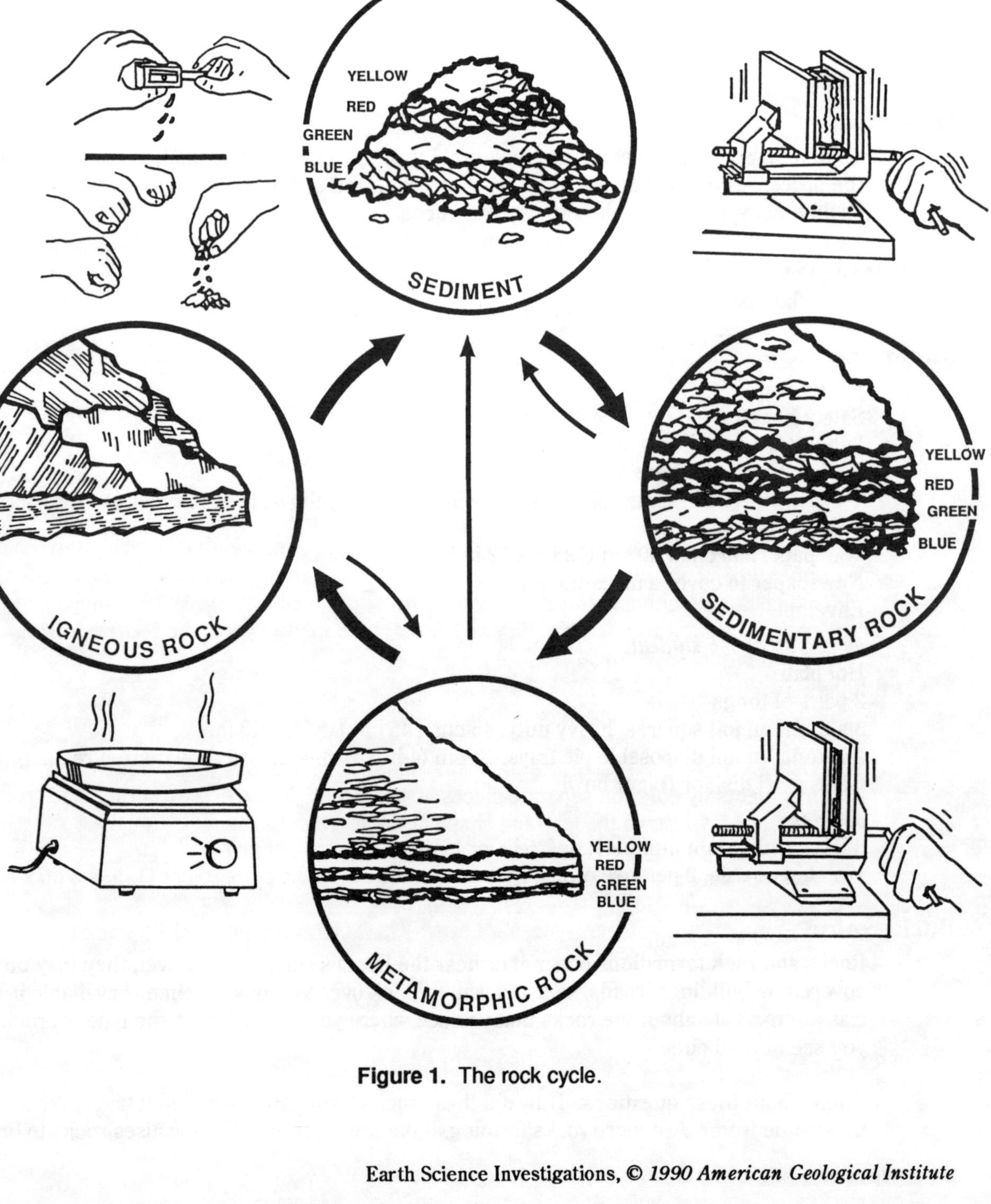

Figure 1. The rock cycle.

Procedure

Students should work individually and in groups of four to six students for each part of this investigation.

CAUTION: ***Students should wear safety goggles and laboratory aprons for all five parts of this investigation.***

Part I. Weathering (class 1)

Weathering is the natural process of breaking down rocks. One sees the effects of weathering on rocks at or near the Earth's surface, but weathering can occur at any depth that air or water can reach.

There are two main types of weathering processes. In **mechanical weathering**, a rock is fragmented without changing its mineral composition. In **chemical weathering**, a rock is broken down by chemical reactions that change its mineral composition and physical and chemical properties.

Chemical weathering results from chemical reactions such as **oxidation**, the combining of oxygen with another substance. An example is the formation of rust. When an iron-containing mineral is wet, oxygen can combine with the iron to form iron oxide, or rust. Oxidation breaks down the chemical bonds between iron and other components in the rock, thus weakening the rock.

Another chemical reaction that may break down rocks is **hydrolysis**, the decomposition of a chemical compound by reaction with water. For example, when the mineral feldspar combines with water, some substances are formed that dissolve, and the remaining chemical elements form a new mineral, kaolinite, a common clay.

Mechanical and chemical weathering operate together to break down rock materials. Fragmenting a rock into smaller pieces increases the surface area exposed to chemical attack, thus speeding up chemical weathering. The decay caused by chemical weathering weakens the rock, making it more susceptible to breakage. Climate affects the relative intensities of these two processes.

1. Cover your group work area with newspaper.

2. Using a pencil sharpener, carefully shave four to six crayons of the same color. Pile the fragments by color on separate pieces of wax paper. You are "weathering" the "rock material." Observe the size and shape of the fragments.

3. Answer the Part I questions.

4. Wrap the fragments by color in wax paper packets. Put each packet in an envelope with your name on it.

5. Clean your group work area.

Questions

1. Are all the "weathered rock fragments" in your pile the same size? Why or why not?
2. What have you observed about the size of rock fragments in nature?
3. Where do rock fragments tend to collect?
4. Where are rocks subject to weathering?
5. How do rocks get smaller in size?
6. How might a tree root become an agent of mechanical weathering? Freezing water?
7. How does mechanical weathering differ from chemical weathering? Which type of weathering process does this activity represent?
8. What kind of climate would provide an atmosphere in which chemical weathering could occur rapidly? Why?

Part II. Erosion and deposition (class 2)

Weathering produces loose, fragmented material that is subject to **erosion** — the movement of weathered material. When the material is transported by water, wind, ice, or gravity, it accumulates in layers of loose, unconsolidated form (such as sand or mud), known as **sediment**. The laying down of sediment is called **deposition**. **Stratification** is the resulting arrangement of the layers of sediment.

1. Cover your group work area with newspaper.

2. Fold one sheet of aluminum foil in half and place it on the work area.

3. Open each student's envelope of "rock" fragments, and place each color in a pile accessible to the group.

4. One student carefully transfers one color of "rock" fragments to the center region of the foil, gently spreading them over an area about 8 cm x 8 cm x 1 cm thick (3 in. x 3 in. x 1/2 in.). A second student carefully places another layer of different-colored fragments on top of the first layer. Continue until each student in the group has placed a color layer of fragments on the pile. Examine the pile of "rock" fragments carefully.

5. Answer the Part II questions.

6. Carefully fold the foil over the fragments. Be sure to allow one cm between the fragments and each of the four foil folds (Figure 2). Label the foil package with the group's name.

7. Clean your group work area.

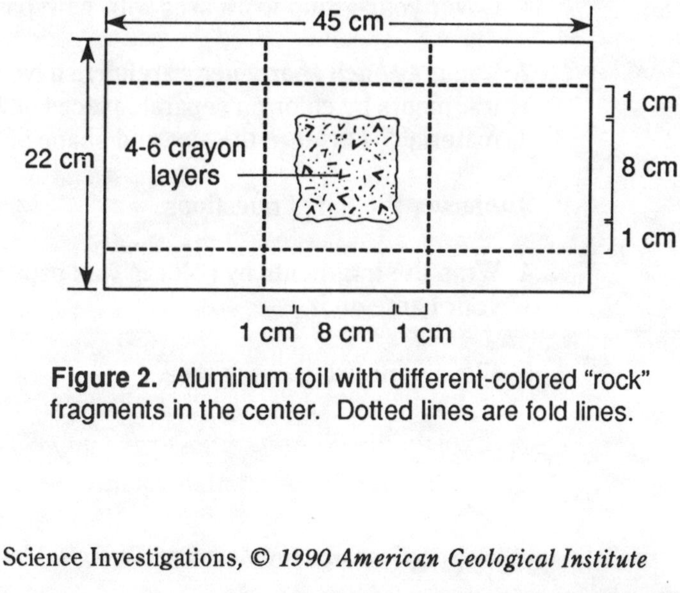

Figure 2. Aluminum foil with different-colored "rock" fragments in the center. Dotted lines are fold lines.

Questions

1. Explain your action as an agent of erosion.
2. Explain the process of erosion.
3. What are some naturally occurring erosional forces (agents)?
4. What are deposition and stratification? What are some conditions that might control deposition?
5. Why are similar-sized rock fragments often found together?
6. Where could you find rock debris fragments in loose layers?

Part III. Sedimentary rock simulation (class 2)

Sedimentary rocks result from the consolidation of sediment or from the chemical precipitation of mineral matter that was once dissolved in water. Sediment is consolidated into rock by compaction and cementation. **Compaction** reduces the size and number of spaces between particles. This compression is produced by the increasing weight of overlying material that is continually being deposited or from pressures produced by Earth movements. **Cementation** occurs when mineral material such as carbonates, silica, and iron oxides chemically precipitate in the spaces between the compacted particles and help to bind the grains together. Minerals that act as such a natural "glue" are called **cementing agents.** Calcite ($CaCO_3$), quartz (SiO_2), and hematite (Fe_2O_3) are common cementing agents.

1. Place the group's folded foil package between two boards, and place this "sandwich" in a vise or two C-clamps. Apply light pressure to the two boards so that the "rock" fragments are slightly compressed.

2. Remove the package from the vise or two C-clamps. Carefully open it and observe the new product. Look specifically at the central region and the outer edges.

3. Carefully break this "sedimentary rock" into two parts and examine the broken edges.

4. Answer the Part III questions.

5. Transfer the loose "rock" fragments and one piece of "sedimentary rock" to an aluminum foil tray. Put your group name on the tray. Return the other piece of "sedimentary rock" to the foil package. Wrap the package as you did in Part II, and place it in an envelope with the group's name on the envelope.

6. Clean your group work area.

Questions

1. Which fragments held together better, those on the outer edges or those in the central region? Why?
2. Describe the layer thickness now compared with its thickness when it was initially deposited on the foil.
3. What happened to the spaces between the "rock" fragments?
4. Explain the difference between compaction and cementation.
5. What are three cementing agents?

Part IV. Metamorphic rock simulation (class 3)

Metamorphic rocks are formed from pre-existing rocks that are changed by heat, pressure, chemical activity, or a combination of the three, without melting. They may become contorted in appearance and actually flow as a plastic material in response to pressures caused by overlying rock load and Earth movements.

1. Cover your group work area with newspaper.
2. Place the group's foil package between the boards and in the vise or two C-clamps, and close the vise or two C-clamps as much as possible. The pressure must be greater than that applied in Part III.
3. Remove the package. Open it, and examine the newly formed "rock."
4. Break the "rock" open, and examine it carefully.
5. Answer the Part IV questions.
6. Place the previously saved "sedimentary rock," some of the "sediment" fragments, and a piece of "metamorphic rock" in a petri dish or finger bowl.
7. Clean your group work area.

Questions

1. Describe the color-layer thickness, comparing "metamorphic rock" with "sedimentary rock."
2. Describe the change in the fragment shape.
3. Why did this change happen?
4. Describe the surface that came in contact with the foil.
5. Why is the surface of the "sedimentary rock" less shiny than the surface of the "metamorphic rock"?

Part V. Igneous rock simulation (classes 3 and 4)

Igneous rocks are formed by the cooling and solidifying of molten or partly molten rock material. When they are formed beneath the Earth's surface, the molten rock, or **magma**, cools slowly, allowing time for mineral crystals to grow to a large size. In contrast, when magma reaches the Earth's surface, such as through volcanic activity, it cools rapidly, which may prevent large crystals from forming.

Section A. (class 3)

CAUTION: *Special concern for safety is essential for this part of the investigation. Students must wear safety goggles throughout this part. They should be especially careful when working with the hot plate and melted wax. Some hot plates may get hot enough to cause burns without showing any signs of being hot.*

1. Cover your group work area with newspaper.

2. Place all crayon fragments in an aluminum foil tray except those that you set aside in Part IV, step 6.

3. Cover the hot plate surface with a square of aluminum foil. Place the tray on the hot plate, and turn the temperature to medium. **CAUTION:** *Avoid dropping wax fragments on the hot plate surface. Do not let the wax heat to the spattering point.*

4. Melt the wax, being careful that the melting process does not occur so rapidly that the molten wax begins to spatter.

5. When most of the "rock" has melted, turn the hot plate off. There should be sufficient heat in the hot plate and in the molten wax to melt the remaining solid mass.

6. Using two pairs of tongs, CAREFULLY lift the tray with the molten wax off the hot plate. Place the tray on several thicknesses of newspaper or on a crushed ice surface until the next day. Both types of areas should be provided for groups to cool their trays. Place your group's name on the tray.

7. Clean your group work area.

Section B. (class 4)

1. Cover your group work area with newspaper.

2. Pick up your tray with the solidified mass, turn it over, and gently tap the bottom so that the wax block falls out. Observe the wax surface that was next to the metal bottom.

3. Compare your group's product with one formed under different temperature conditions (one placed on crushed ice).

4. Break the wax block into several pieces. Place one piece with the most interesting features in the aluminum foil tray with the other "rocks." Return the remaining fragments to the melting tray.

5. Again, cover the plate surface with a square of foil. Then, place the aluminum tray on the hot plate, and turn the temperature to medium. Melt the wax. **CAUTION:** *Remember to turn off the hot plate before all the wax is melted. Do not let the wax spatter.*

6. While the wax is still in the molten state, using two pairs of tongs, a student from each group should CAREFULLY pour the melted wax into a tray of ice water. The "igneous rock" formations will solidify in a short period of time.

7. Answer the Part V questions.

8. Clean your group work area.

Questions

1. What happened to the "rock" fragments (both loose and in a compacted state) when they were heated on the hot plate?
2. Where would this melting activity happen in the Earth?

3. What did you observe on the bottom of the "igneous rock" when it cooled the first time?
4. How was the appearance different between the wax cooled on the newspaper and the wax cooled on the crushed ice?
5. Why do you think the colored crystals were found in the bottom of the wax mass?
6. Explain what happened to the wax when you poured it into the cold water.

Additional resources

The Rock Cycle. AGI/EBE Earth Science Programs. Chicago: Encyclopaedia Brittannica Educational Corp., 1982. Film: 16 mm, 22 min., sound.

Answer Key

Questions

Part I.

1. The fragments should display a variety of fragment shapes, but the thickness will be somewhat consistent with the pencil sharpener used.

2. They tend to vary widely in size in any given location.

3. As rock fragments form, they seek the lowest possible level.

4. Rocks at the surface of the Earth and within the first few meters of the surface are exposed to mechanical and chemical weathering.

5. Rock fragments are broken down by mechanical and chemical weathering.

6. A tree or plant root growing in a small crack in a rock exerts pressure that enlarges the crack, splitting the rock apart. Water freezes and expands in small cracks, forcing the rock material to separate.

7. Mechanical weathering does not result in a new product. Chemical weathering results in new products, and both the physical and chemical properties are altered. This activity represents mechanical weathering.

8. In hot, humid climates, the conditions for oxidation and hydrolysis are optimum for rapid decomposition of rocks.

Part II.

1. Students picked up "rock" fragments and physically moved them to form the colored layers.

2. Erosion is the movement of particles by a variety of forces.

3. Erosional forces include wind, water, ice (glaciers), and gravity (landslides).

4. Deposition is the laying down of transported rock material. The positioning of the material in layers is stratification. Factors that may control the layering include the rate of wind or water movement and the density of materials being moved by the erosional force.

5. Water and wind sort sediment by selectively moving or depositing material depending on the material's size and/or density. The ability of these transporting agents to move material is affected by their velocity. For example, along a small, shallow, fast-moving creek without meanders, smaller particles will be along each bank where the water velocity is slower, and larger particles will be in the middle of the creek where the water velocity is greater.

6. Sand dunes, beaches, alluvial fans, deltas, and landslides are places where one could find rock fragments in loose layers.

Part III.

1. Crayon material was squeezed outward from the central region. Fragments nearer the center were compacted more than those towards the edges, and they held together better.

2. The color-layer thickness has been significantly reduced due to compaction.

3. They have been reduced in size because of the pressure applied by the vise or C-clamps.

Answer Key

4. Compaction is a process in which particles are forced together and the space between them is reduced. This compression results from the weight of overlying materials or from pressure of Earth movements. The vise or pair of C-clamps was the compacting force in this investigation, simulating pressure on materials in the Earth's crust placed by overlying materials. Cementation is a process in which mineral material forms in the spaces between particles, binding the particles together.

5. Calcite ($CaCO_3$), quartz (SiO_2), and hematite (Fe_2O_3) are cementing agents.

Part IV.

1. As a result of being under increased pressure, the color layers in the "metamorphic rock" are compacted into thinner bands of color than those in the "sedimentary rock."

2. Fragment shape was originally angular, but as pressure increased, the fragments began to fuse and lose their distinct shapes.

3. Under increased pressure, the fragments lost their rigidity; they became more pliable and began to fuse with neighboring fragments.

4. This "rock" surface took the same surface texture as that of the smooth foil surface. It, too, became a smooth, shiny surface.

5. Fragments in the "sedimentary rock" have been under less pressure (not as close together) than those in the "metamorphic rock." Therefore, they have not formed such a smooth outer surface. They reflect less light. The "metamorphic rock" particles are closer together, resulting in a surface that reflects more light.

Part V.

1. The fragments melted.

2. While volcanic activity appears to be a surface phenomenon, since that is what is seen, heat and pressure cause Earth material to melt deep below the surface, possibly as deep as 250 km (155 miles).

3. Students should have observed a series of colored crystal-like bodies. These always seem to appear only on the bottom of a solidified wax body.

4. Wax that cools slowly (on newspaper) almost always forms larger masses of crystal-like bodies than the wax that cools quickly (on crushed ice). This phenomenon in the simulation parallels crystal growth in the natural world. Note: The brand of crayon seems to influence the size of the crystals that form. This effect may be due to the pigment in the wax base from which the crayons are formed.

5. Pigment density may play a role, which parallels the role that chemical density and heat of fusion play in the formation of crystalline substances in rocks and minerals.

6. The outer surface came into contact with the cold water and began to solidify before the interior. It took on a smooth, "pillowed" appearance. This appearance parallels a formation called pillow lava, which results when 600+°C magma comes in contact with cold water on the ocean floor.

Determining the Density of Soil Particles

DONATO DIGREGORIO
Jefferson Junior High School
Oak Ridge, Tennessee

Objective

To determine the mass, volume, and percent of air space of soil samples so that the particle density of soils can be calculated.

Time required

Two class periods. Allow more time if using more than two samples.

About three to four weeks before using this investigation, the teacher or students should collect several soil samples (at least two different kinds). A small shovelful of each will be enough for a class. Spread the samples on newspapers to dry. When the samples are dry, grind them to remove lumps.

Materials

per group of 2-4 students

Dry soil samples: about 35 ml each of at least two different kinds of soil. Number each sample.
2 pieces of paper: one should be 10 cm x 10 cm (4 in. x 4 in.), one 8 1/2 in. x 11 in. (notebook paper).
Balance
50-ml graduated cylinder
Long-stemmed funnel
Water: 30 ml for each soil sample. Any amount from 25 to 40 ml works well; however, it is critical that students record the actual amount used.
Wooden or metal stirring rod
Container for discarded samples
Calculator (optional)

Background

Density, or mass per unit volume, is a physical property that earth scientists use to compare Earth materials. **Mass** is the amount of matter a substance contains. **Volume** is the amount of space a substance occupies. The density of regularly shaped objects such as cubes, spheres, and pyramids can easily be found by using appropriate mathematical formulae to calculate the volume of the object and by determining the mass of the object on a balance. Then the density formula shown below is applied.

$$\text{Density} = \frac{\text{Mass}}{\text{Volume}} \qquad \text{or} \qquad D = \frac{M}{V}$$

Similarly, the density of an irregularly shaped object such as a stone can be found by measuring the mass, determining the volume of the object by water displacement, and applying the density formula.

Somewhat more of a problem, however, is encountered when determining the density of granular solids such as sand or soil. These types of solids contain air space, which must be allowed for when determining the volume.

This investigation shows how to determine the density of soil particles so that air space is taken into account. You will be able to compute the soil particle density after you have measured the mass, volume, and percent of air space. The density of a solid is often expressed in grams (mass) per cubic centimeter (volume). Your density calculations will be in grams per milliliter. One milliliter equals one cubic centimeter.

Procedure

Use the following procedure for determining the particle density of each soil sample.

CAUTION: *Dry soils can be very dusty. Handle them with care to keep down the amount of dust in the air. If the air is dusty, it may be advisable to wear goggles and/or a mask over your nose and mouth.*

Part I. Measuring the volume (with air space)

Pour a soil sample into a graduated cylinder, and record the volume below. You should have 35 ml of dry soil per sample. The volume of dry soil is considered to be the volume of soil with air space (S + A).

	Sample number			
	1	**2**	**3**	**4**
Volume of dry soil (ml)	....	....	....	....

Part II. Determining the mass

1. Place a paper square on a balance, and measure its mass. Record the mass on the next page.

2 Place the dry soil sample (35 ml) on the paper, and measure and record the combined mass of the paper and soil.

3. Calculate the mass of the soil: subtract the mass of the paper from the combined mass of the paper and soil. Record your answer in grams (g).

	Sample number 1	2	3	4
Mass of paper and soil (g)	....	....	....	....
Mass of paper (g)	....	....	....	....
Mass of soil (g)	....	....	....	....

Part III. Determining the volume (without air space)*

1. Using a long-stemmed funnel, pour 30 ml of water (W) into the graduated cylinder so that the cylinder sides remain dry. Pour the water slowly and carefully. If the sides get wet, soil sticking to them may affect your readings.

2. Roll a piece of notebook paper to make a funnel. Insert the paper funnel midway between the top of the cylinder and the top of the water. Slowly pour the soil (S) through the funnel, making sure that none sticks to the sides of the cylinder. It is important to pour the soil slowly so as not to create trapped air pockets. Be careful to keep moving the funnel up as the water level in the cylinder rises.

3. Using a stirring rod, stir the water and soil gently to eliminate any air pockets. Record the level, or volume, of the water after the soil has been added (W + S).

4. Calculate the volume of the soil without air space: subtract the level of the water (W) from the level of the water and soil (W + S). Use the formula (W + S) - W = S. Record your answer in milliliters (ml).

	Sample number 1	2	3	4
Volume of water and soil (ml)	....	....	....	....
Volume of water (ml)	....	....	....	....
Volume of soil without air space (ml)	....	....	....	....

Part IV. Determining the percent of air space*

In Part I, you measured the volume of dry soil, which is considered to be the volume of soil with air space (S + A). In Part III, you calculated the volume of soil alone, without air space (S).

1. Calculate the volume of air space (A): subtract the volume of soil without air space from the volume of soil with air space. Use the formula (S + A) - S = A. Record your data and answer on the next page.

---

*The procedures for determining the volume of soil without air space and the percent of air space in the soil have been adapted from Haber-Schaim et al. (1982, 25-26; *Teacher's Guide*, 1982, 30-31).

2. Calculate the percent of air space: divide the volume of air space (A) by the volume of dry soil (S + A), and multiply your answer by 100. Use the formula below, and record your answer.

$$\frac{A}{(S + A)} \times 100 = \text{percent of air space}$$

	Sample number 1	2	3	4
Volume of dry soil (ml)	....	....	....	....
Volume of soil alone, without air space (ml)	....	....	....	....
Volume of air space (ml)	....	....	....	....
Percent of air space (%)	....	....	....	....

Part V. Determining the density of soil particles

1. Calculate the density by using the density formula: density equals mass of soil divided by the volume of soil without air space. Use the mass (in grams) calculated in Part II and the volume (in milliliters) calculated in Part III.

2. Fill in Table 1 on the worksheet.

	Sample number 1	2	3	4
Particle density (g/ml)	....	....	....	....

Part VI. (Optional)

Each group needs one copy of Table 2 on the worksheet for each soil sample. The groups complete the table by sharing data. Variations in measurements provide an opportunity for class discussion of experimental error and standard error of measurement.

Questions

1. Is it possible for any two soil samples to be composed of exactly the same materials in exactly the same amounts? Explain your answer.
2. How do the size, shape, and arrangement of particles in your soil samples affect the density measurements?

Additional resources

Haber-Schaim, Uri; Abegg, Gerald L.; Dodge, John H.; and Walter, James A. *Introductory Physical Science.* 4th ed. Englewood Cliffs, N.J.: Prentice Hall, 1982, p. 25-26.

Haber-Schaim, Uri; Abegg, Gerald L.; Dodge, John H.; and Walter, James A. *Introductory Physical Science: Teacher's Guide and Resource Book.* 4th ed. Englewood Cliffs, N.J.: Prentice Hall, 1982, p. 30-31.

Name ....

Class ....

Date ....

Determining the Density of Soil Particles

Worksheet

Table 1
Summary of group data.

	Soil sample			
	1	**2**	**3**	**4**
Mass (g)	....	....	....	....
Volume (ml)	....	....	....	....
Air space (%)	....	....	....	....
Particle density (g/ml)	....	....	....	....

Table 2
Summary of class data for soil sample number ....

Student group	Mass (g)	Volume (ml)	Air space (%)	Particle density (g/ml)
1	....	....	....	....
2	....	....	....	....
3	....	....	....	....
4	....	....	....	....
5	....	....	....	....
6	....	....	....	....
7	....	....	....	....
8	....	....	....	....
9	....	....	....	....
10	....	....	....	....
11	....	....	....	....
12	....	....	....	....
13	....	....	....	....
14	....	....	....	....
15	....	....	....	....
Average	....	....	....	....

Answer ✓ Key

Questions

1. Soil is a mixture of weathered rock material that usually contains organic matter. The physical property studied in this investigation is particle density. If the densities of the samples are exactly the same, students might conclude that the soil samples were the same. It would certainly be coincidental if the samples were composed of exactly the same materials in exactly the same amounts. Numerous other tests besides particle density would be needed to determine the composition of the soil. Perhaps the most that could be concluded is simply that the soils have similar or different particle densities.

   In most soils, the average density of particles is about 2.65 g/ml because the minerals most prevalent in soils (quartz, feldspar, clay) have density values generally in the range of 2.6 to 2.7 g/ml. The presence of minerals high in iron content increases the average density of particles, while the presence of organic matter lowers the density.

   Student groups may obtain different particle densities for the same sample. Such variations should encourage discussion of possible sources of measurement error.

2. This investigation centers around determining the density of soil particles. That density is affected only by the composition of the soil, because the air space has been excluded from the volume in the calculation.

   The size, shape, and arrangement of particles in the soil samples affect the percent of air space, or porosity. Soils with rounded particles such as sand do not pack well. They tend to have a greater porosity than soils with angular particles. Porosity does affect the density of the whole soil, that is soil with air. This density is known as the bulk density. The bulk density can be determined by dividing the mass of the soil (Part II) by the volume of the dry soil with air space (Part I).

Analyzing a Rock Containing Two Minerals

FRANK H. ROBERTS

Delaware County Christian School
Newtown Square, Pennsylvania
Montgomery County Community College
Blue Bell, Pennsylvania

Objective

To find out the percentages of two minerals in a rock.

Time required

Two class periods: one double period if a centrifuge is used or two separate periods to allow time for the insoluble material to settle.

Materials

per class
Concentrated hydrochloric acid to be dispensed by the teacher
Balances that read to 0.01 gram
Centrifuge (optional)
Goggles to protect eyes while heating acid
Aprons

per student
Crushed specimens of a rock: siliceous limestone or calcareous sandstone
Test tube
Test tube clamp
Test tube rack
Bunsen burner
Hydrochloric acid: 10% solution in a small dropping bottle or plastic squeeze bottle

Background

Most rocks contain more than one mineral. For economically valuable rocks, it is often useful to know how much of each mineral is present in the rock. A typical example is limestone, which is mostly calcite (calcium carbonate) or dolomite (calcium/magnesium carbonate). Limestone or dolomite usually also contains a certain amount of quartz (silicon dioxide) or other minerals. In this investigation, you will remove calcite ($CaCO_3$) from a rock by treating the rock with acid. This procedure will dissolve the calcite but leave the quartz (SiO_2) behind. Obtaining the mass of the rock before and after acid treatment enables you to determine the percentages of calcite and quartz in the rock.

Procedure

CAUTION: *Concentrated hydrochloric acid is corrosive to eyes and skin. Even dilute solutions and vapors may be dangerous. To dilute, always add the acid to water, never water to the acid. Students should wear safety goggles and aprons when working with hydrochloric acid. Use a test tube clamp to hold a test tube containing the hydrochloric acid.*

1. Place a test tube on a balance, and record its mass to the nearest hundredth of a gram (for example, 7.43 g). Write your data in Table 1.

2. Place one to two grams of crushed rock in the test tube, and measure the mass of the test tube with the rock in it. Set it in the test tube rack, and record the mass in the table.

Table 1
Experimental data.

Rock specimen name	....
Mass of test tube (g)	....
Mass of test tube and rock (g)	....
Mass of test tube and quartz (g)	....

3. Obtain a small bottle of 10% hydrochloric acid from your teacher, and add the acid to the crushed rock in the test tube a drop or two at a time. **CAUTION:** *Leave the test tube in the rack while the acid reaction takes place. Do not allow the liquid to foam up and run out of the test tube.*

4. When the bubbling has stopped or the test tube is about half filled with acid, ask the teacher to add a drop of concentrated hydrochloric acid. Wait until the bubbling stops and add another drop. Repeat this step until no reaction occurs when a drop of concentrated acid is added. The bubbling indicates that the calcite is dissolving.

5. Use a centrifuge to bring the insoluble material to the bottom of the tube. If a centrifuge is not available, put the test tube, labeled with your name, in a test tube rack until the next class period.

6. When the insoluble material in the test tube has settled to the bottom of the liquid, pour off the clear liquid into the sink, being careful not to allow any of the solids at the bottom to be washed out with the liquid. The dissolved calcite has been washed away by the acid, and now only the quartz part of the rock remains, with some water and a trace of acid.

7. What remains in the test tube must be dried. (The water in the test tube probably weighs more than the whole rock did at the beginning.) Using the test tube clamp, hold the test tube almost horizontally, and carefully tap the material to one side, away from the bottom of the tube. Still holding the test tube almost horizontally, carefully heat it over a burner. Heat the tube slowly, removing it from the flame each time any boiling starts. **CAUTION:** *When heating a test tube, always hold it with a test tube clamp. Do not heat the bottom of the test tube when holding the tube vertically. Such action may result in expelling the contents onto the table. Work in a well-ventilated area, and avoid inhaling vapors given off by the acid.*

8. When the test tube and its contents are completely dry, measure and record its mass. This mass is that of the test tube and the quartz.

Questions

1. What is the mass of the original rock containing calcite and quartz? Subtract the mass of the test tube from the mass of the test tube and rock.
2. What is the mass of the calcite in the rock? Subtract the mass of the test tube and quartz from the mass of the test tube and rock.
3. What is the percent of calcite in the rock? Divide the mass of the calcite by the mass of the original rock. Multiply that answer by 100 to get a percent.
4. What is the mass of the quartz in the rock? Subtract the mass of the test tube from the mass of the test tube and quartz.
5. What is the percent of quartz in the rock? Divide the mass of the quartz by the mass of the original rock. Multiply that answer by 100 to get a percent.

Additional resources

Compton, Robert R. *Manual of Field Geology.* New York: John Wiley and Sons, 1962, p. 217.

Huang, Walter T. *Petrology.* New York: McGraw-Hill Book Co., 1962, p. 214, 240.

Pettijohn, F.J. *Sedimentary Rocks.* New York: Harper and Row, Publishers, 1957, p. 381-387, 651-660.

Weller, J. Marvin. *Stratigraphic Principles and Practice.* New York: Harper and Row, Publishers, 1960, p. 93, 118-120, 294.

Answer Key

Procedure

This investigation requires samples of an impure calcareous rock. A pure limestone cannot be used because it will dissolve completely. The author brought a chunk of rock from a local outcrop into the classroom, hammered a little off, and crushed it. **CAUTION:** *Before hammering a rock in the classroom, cover the rock with a heavy cloth such as burlap to contain flying chips.*

If suitable local rock is not available, you may be able to purchase limestone chips used for landscaping.

Be sure to do a pre-lab run-through with a sample of the specific rock you plan to use in the classroom.

Results of this experiment will depend on the rock chosen for analysis. A relatively pure limestone may have less than 3% insoluble material. Calcareous sandstone may have only 30% soluble carbonate.

Comparing Water Hardness

Cora M. Turpin
Philadelphia High School for Girls
Philadelphia, Pennsylvania

Objective

To investigate the hardness of various samples of water.

Time required

One class period
More time may be required, depending on the number of water samples used.

Materials

per group of 2 students

Water samples: 50 ml of different kinds of water

- **CAUTION:** *Students should be careful when working with water such as pond or mud puddle water, which may be contaminated.*
- Distilled water (This will be the control.)
- Bottled commercial water
- Mineral water
- Mud puddle water
- Pond water
- Rain water
- Sea or salt water
- Spring water
- Tap water
- Well water
- Other samples as available

Test tubes: one for each water sample
Sterile rubber corks: one for each test tube
Eyedropper
Test tube racks
Labels and tape to mark the water samples
Graduated cylinder
Container of liquid hand soap. Do not use detergent; it will lather in hard or soft water.
Watch with second hand

per student

Graph paper

Background

Note: Students should have a general knowledge of ions, anions and chemical precipitation.

When soap is added to hard water, curd forms that you recognize as a greasy ring around a sink or bathtub. Few suds form because the water contains dissolved magnesium, ferrous, and/or calcium ions that combine with the soap to form the insoluble residue. To get a lasting lather, all of the curd-forming ions must be removed by continuously adding soap. The more soap that is needed to lather, the harder the water.

Procedure

This investigation follows the common practice of expressing liquid measurements in milliliters (ml). If your graduated cylinder is marked in cubic centimeters (cc), one cc equals one ml.

1. As you work with each sample, record your data in Table 1 on the worksheet.

2. Be certain all test tubes are clean and labeled.

3. Using a graduated cylinder, put 50 ml of each water sample in a test tube. Be careful not to contaminate your samples!

4. Using an eyedropper, add liquid soap, drop by drop, to the distilled water (control water sample). Seal with a cork. Shake the sample for 10 seconds after each drop until the sample soap bubbles last two minutes. In Table 1, record the number of drops needed to make the soap bubbles last two minutes. This will be the least amount of soap you will need to make soap bubbles in your samples.

5. Repeat steps 3 and 4 for each sample. Use your control to estimate the number of drops needed to make soap bubbles in each of the samples.

6. Show your data on graph paper: the x-axis (horizontal) should show the sample, and the y-axis (vertical) the number of drops of soap needed for bubbles to last for two minutes.

7. Devise a ranking scale for expressing the hardness of water, and rank the water samples.

Questions

1. Why must all equipment be clean?
2. How does soap differ from detergent?
3. Why can't detergent be used for this experiment?
4. Why is distilled water the control sample?
5. What information does making soap bubbles in distilled water give for this experiment?
6. Is 50 ml of each sample the best amount of water to be used for this experiment? Could it be more? Less? Why or why not?
7. Each sample requires 50 ml (amount) to be shaken for 10 seconds (time), and the bubbles must last two minutes (evidence). What is the variable?
8. From your data, calculate the amount of soap needed to make lather in 10 ml of each sample and two liters of each sample.

9. Would a different kind of soap give the same results?
10. Write three ideas that you learned from doing this experiment.
11. Write three questions that you have as a result of this experiment.
12. (Optional) If approximately seven drops of liquid soap equal one ml, calculate (a) the volume of soap in one drop, and (b) the mass of one drop of soap. Assume the density of the soap is 1.21 g/ml.
13. (Optional) How many moles of soap are needed for soft water if the volume of water is 50 ml and the ratio of soap to water is two moles of soap (molecular weight = 306 g/mole) to three moles of water (molecular weight = 18 g/mole)? Note: density of water = 0.98 g/ml.

Worksheet

Name ....

Class ....

Date ....

Table 1
Experimental data.

Type of water sample	No. of drops of soap	Length of time bubbles lasted
....	....	....
....	....	....
....	....	....
....	....	....
....	....	....
....	....	....
....	....	....
....	....	....
....	....	....
....	....	....

Answer Key

Questions

1. Equipment must be clean so that contaminants from the glassware do not affect experimental results.

2. Soap lathers in soft water only. Detergents lather in hard or soft water.

3. Because detergents lather in hard water and soft water, no scale can be devised.

4. Distilled water is pure water with no mineral content.

5. Bubbling or forming a lather in distilled water gives the experimenter an example of the effect to expect and establishes the minimum number of soap drops that will be needed to make a lather in 50 ml of each sample.

6. This is a debatable thought question. Look for logic in students' answers.

7. The variable is the amount of soap used.

8. The results should be consistent with the data.

9. Predictably, changing the variable will vary the results.

10. and 11. Answers will vary.

12. a. Volume of soap in one drop:
    1 drop x (1 ml/7 drops) = 0.14 ml

    b. Mass of one drop of soap:
    0.14 ml x (1.21 g/ml) = 0.17 g

13. Moles of soap needed:

$$50 \text{ ml water} \times \frac{0.98 \text{ g water}}{1 \text{ ml water}} = 49.0 \text{ g water}$$

$$49.0 \text{ g water} \times \frac{1 \text{ mole water}}{18 \text{ g water}} = 2.72 \text{ moles water}$$

$$2.72 \text{ moles water} \times \frac{2 \text{ moles soap}}{3 \text{ moles water}} = 1.81 \text{ moles soap}$$

Determining Ground-Water Contamination

Paul L. Garvin
Cornell College
Mount Vernon, Iowa

Objective

To determine, by preparing and analyzing contour maps, which way a plume of contaminated ground water will move, which drinking water wells will be affected, and how long it will take the contaminants to reach the wells.

Time required

Two to three class periods

Materials

per student
2 copies of Worksheet 2
Straight edge
3 pencils, different colors
Calculator
2 sheets of tracing paper (optional): 8 1/2 in. x 11 in.

Background

Note: Students should know basic procedures of contouring manually or by computer.

Students should understand the following terms (Figure 1):

ground water Water that accumulates beneath the ground surface and fills the available pore spaces.

water table The boundary between the area saturated with water (zone of saturation) and the overlying area containing air and water (zone of aeration).

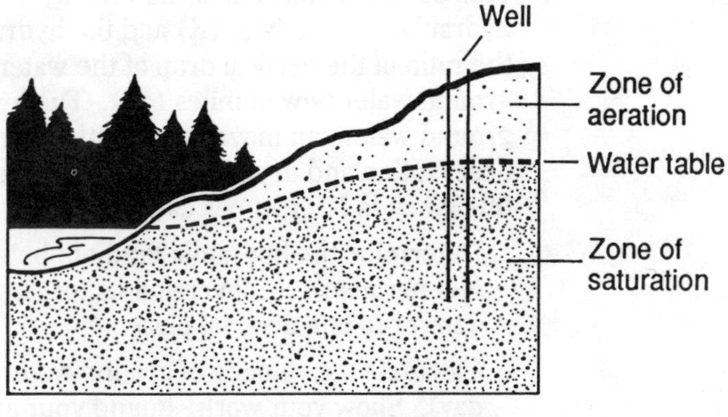

Figure 1. Ground-water terminology.

Procedure

Underlying a military base in northeastern Michigan is a shallow sand and gravel **aquifer**, a subsurface layer that is permeable enough to conduct ground water and to yield water readily to wells and springs. The water table lies between 10 and 25 feet below the ground surface. A leak in a buried storage drum has allowed a toxic organic liquid to enter the aquifer. This contamination is a potential threat to drinking water supplies on the base.

Table 1 on Worksheet 1 lists the ground-surface elevation and depth to the water table for 55 wells on the military base. English units are used rather than metric units, because ground elevation data are given in feet. Conversion to metric units gives fractional elevation data that are more difficult to use. The wells were drilled for various purposes, and their locations are shown on the map of the base (Figure 2) on Worksheet 2.

1. Using the data in Table 1, calculate the elevation of the water table at each well by subtracting the depth to the water table from the elevation of the well at the ground surface. Record your answers in Table 1.

2. On the map on one copy of Worksheet 2 or on a tracing overlay, plot the elevation of the water table at each well.

3. Using one of the colored pencils, contour the water table elevation on the map or overlay. Use a contour interval of one foot. The contour lines you have drawn are called **equipotential lines** and show the general location of the water table.

4. The direction of ground-water flow is generally perpendicular to the equipotential lines, moving from higher to lower elevations. Using a second color, draw arrows (flow lines) at several places on the map or overlay to show the directions of ground-water movement.

Questions

1. Based on the direction of ground-water movement, which of the drinking water wells is likely to be contaminated by **effluent**, or outflow, from the leaking storage drum?

   The pollution plume will spread quickly to a width of about 500 feet in the vicinity of the well. Shade in the area of the pollution plume.

2. The velocity of ground-water movement can be determined from Darcy's law: $V = K\,\Delta H/\Delta L$. This equation shows that the velocity (V) of ground-water movement is a product of the hydraulic conductivity (K) and the hydraulic gradient ($\Delta H/\Delta L$). The hydraulic gradient is the ratio of the vertical drop of the water table in feet (ΔH) to the horizontal distance of ground-water flow in miles (ΔL). The hydraulic conductivity describes the rate at which ground water can move through the aquifer. It is determined by experiment and observation and depends on aquifer permeability and fluid properties. For the aquifer in this lab, it is 100 feet per day.

   a. Determine the hydraulic gradient between the storage drum and the threatened well (in feet per mile). (Conversion factor: 1 mile = 5,280 feet.) Show your work. Round your answer to one decimal place.

   b. Calculate the velocity of ground-water flow from the storage drum to the well (in feet per day). Show your work. Round your answer to one decimal place.

Name ....

Class ....

Date ....

Worksheet 1

Table 1

Ground-surface elevations and water-table depths for selected wells at the military base.

Well number	Elevation of well (ft.*)	Depth to water table (ft.)	Elevation of water table (ft.)	Well number	Elevation of well (ft.*)	Depth to water table (ft.)	Elevation of water table (ft.)
AF 2	613	24	....	H 10	619	19	....
AF 3	616	25	....	H 11	618	19	....
AF 4	614	25	....	H 13	618	19	....
AF 5	611	22	....	H 14	618	19	....
AF 18	617	18	....	O 5	616	19	....
AF 52	611	20	....	O 6	615	23	....
AF 61	619	21	....	O 8	615	19	....
AF 62	613	20	....	O 9	611	19	....
AF 64	611	20	....	R 3	609	21	....
AF 70	615	18	....	R 4	612	23	....
AF 72	615	18	....	R 5	615	22	....
AF 74	615	20	....	R 6	617	24	....
AF 75	615	20	....	R 7	617	22	....
AF 76	614	21	....	R 8	616	20	....
G 7	619	19	....	R 11	615	22	....
G 8	616	24	....	R 17	617	23	....
G 9	609	21	....	R 18	617	23	....
G 10	615	26	....	R 21	618	20	....
G 11	608	20	....	R 23	617	22	....
G 12	614	23	....	R 25	613	21	....
G 17	618	22	....	R 76	613	19	....
G 20	615	20	....	R 77	613	20	....
H 1	621	17	....	R 78	608	20	....
H 2	621	17	....	R 79	614	24	....
H 3	621	19	....	R 80	614	25	....
H 4	621	19	....	R 84	613	21	....
H 5	621	20	....	Y 5	608	19	....
H 8	618	18	....				

Source: *U.S. Geological Survey Water Resources Investigations Report 83-4002* (1983). Data were slightly modified to simplify the investigation.

*Elevations in feet above mean sea level.

c. Using the formula, time = distance/velocity, determine how long it will take the contaminants to reach the well. (Assume no loss of contaminants by absorption.) Give your answer in years, and show your work. Round your answer to one decimal place.

3. (Optional) A well is scheduled to be drilled at P (Figure 2). Because of the high pumping requirement, the water table at the well is expected to be lowered by 15 feet. Table 2 shows water-table lowerings within the **cone of depression**, the area around the well where the water table will be affected. The cone of depression will extend to a radius of 1,000 feet around the well.

Table 2
Water-table lowerings around well P.

Distance from well (ft.)	Average amount of water-table lowering (ft.)
0	15
250	10
500	6
750	3
1,000	1

a. On the other copy of Worksheet 2 or on another tracing overlay, use a third color to recontour that portion of the water-table surface affected by the new well. Show with arrows the direction of ground-water flow. Use a contour interval of two feet.

b. What effect will the new well have on the direction of movement of contaminants from the storage drum? What are the consequences of the change?

4. What have you learned from this exercise about ground-water movement and waste disposal?

Name ....

Class ....

Date ....

Determining Ground-Water Contamination

Worksheet 2

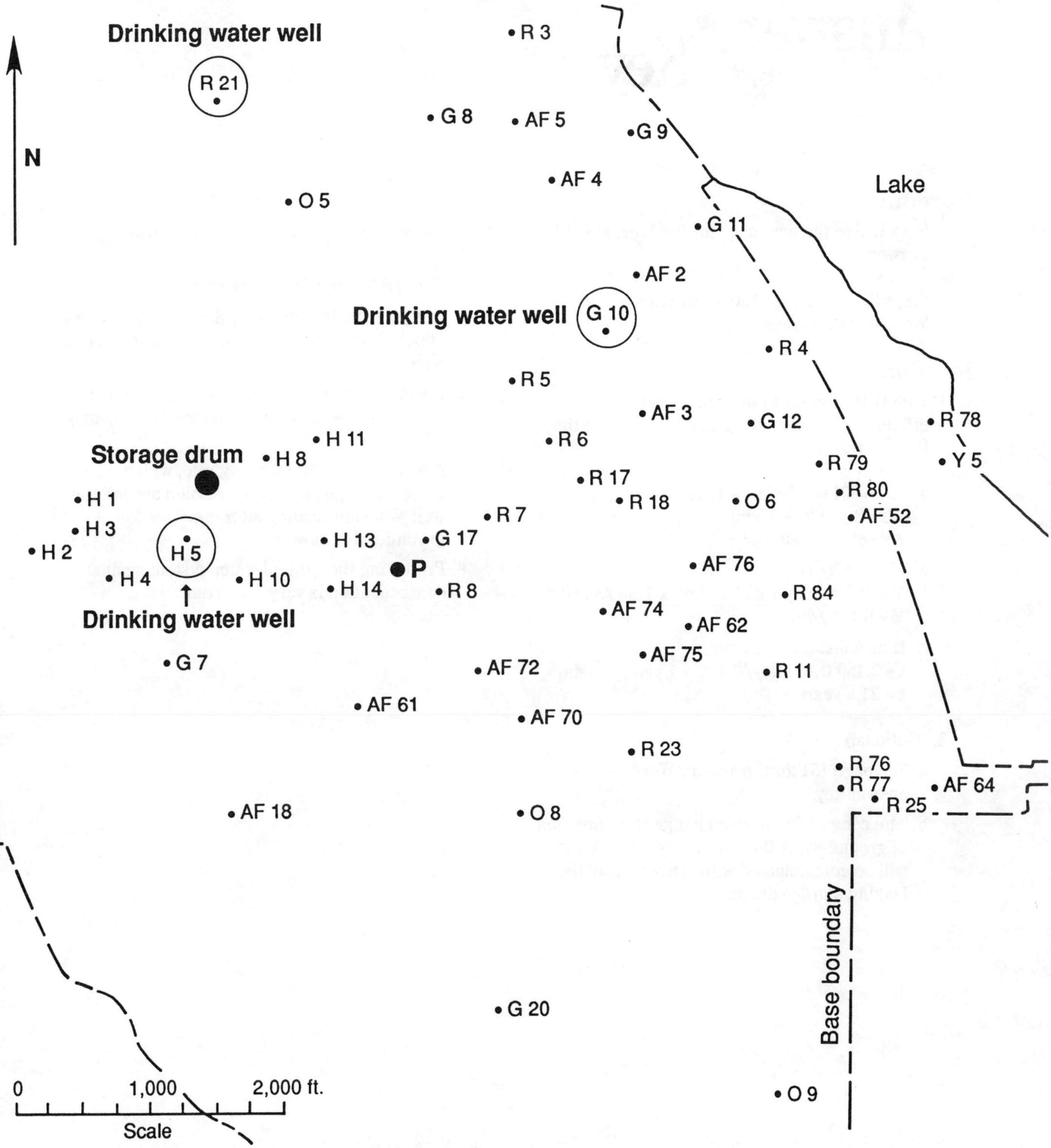

Figure 2. Map of military base showing locations of selected wells; adapted from *U.S. Geological Survey Water Resources Investigations Report 83-4002* (1983).

Answer Key

Procedure

Step 1. See the next page for the Worksheet 1 answer key.

Steps 2-4. See page 150 for the first Worksheet 2 answer key.

Questions

1. Well G 10 is likely to be contaminated by effluent. The pollution plume is shown on the first Worksheet 2 answer key.

2. a. ΔH = 601 ft. - 589 ft. = 12 ft.
   ΔL = 3,150 ft. = 0.6 mi.
   $\Delta H/\Delta L$ = 20.0 ft./mi.

   b. $V = K\,\Delta H/\Delta L$
   V = 100 ft./day x 20.0 ft./mi. x 1 mi./5,280 ft.
   V = 0.4 ft./day

   c. time = distance/velocity
   t = 3,150 ft. x 1 day/0.4 ft. x 1 year/365 days
   t = 21.6 years

3. (Optional)
   a. See page 151 for the second Worksheet 2 answer key.
   b. The cone of depression changes the direction of ground-water flow toward well P. Well P will be contaminated with effluent from the leaking storage drum.

4. Answers will vary. Students should have learned the following:
   a. Ground water moves very slowly.
   b. It is possible to determine direction and speed of pollution migration using a small amount of data.
   c. A well near the contamination source may not be affected, while one much farther away may be.
   d. A well with a high pumping rate, which creates a large cone of depression around the well, will significantly alter the direction of ground-water flow.
   e. Protecting the ground-water system against contamination is very important.

Table 1
Ground-surface elevations and water-table depths for selected wells at the military base.

Well number	Elevation of well (ft.*)	Depth to water table (ft.)	Elevation of water table (ft.)	Well number	Elevation of well (ft.*)	Depth to water table (ft.)	Elevation of water table (ft.)
AF 2	613	24	589	H 10	619	19	600
AF 3	616	25	591	H 11	618	19	599
AF 4	614	25	589	H 13	618	19	599
AF 5	611	22	589	H 14	618	19	599
AF 18	617	18	599	O 5	616	19	597
AF 52	611	20	591	O 6	615	23	592
AF 61	619	21	598	O 8	615	19	596
AF 62	613	20	593	O 9	611	19	592
AF 64	611	20	591	R 3	609	21	588
AF 70	615	18	597	R 4	612	23	589
AF 72	615	18	597	R 5	615	22	593
AF 74	615	20	595	R 6	617	24	593
AF 75	615	20	595	R 7	617	22	595
AF 76	614	21	593	R 8	616	20	596
G 7	619	19	600	R 11	615	22	593
G 8	616	24	592	R 17	617	23	594
G 9	609	21	588	R 18	617	23	594
G 10	615	26	589	R 21	618	20	598
G 11	608	20	588	R 23	617	22	595
G 12	614	23	591	R 25	613	21	592
G 17	618	22	596	R 76	613	19	594
G 20	615	20	595	R 77	613	20	593
H 1	621	17	604	R 78	608	20	588
H 2	621	17	604	R 79	614	24	590
H 3	621	19	602	R 80	614	25	589
H 4	621	19	602	R 84	613	21	592
H 5	621	20	601	Y 5	608	19	589
H 8	618	18	600				

Source: *U.S. Geological Survey Water Resources Investigations Report 83-4002* (1983). Data were slightly modified to simplify the investigation.

*Elevations in feet above mean sea level.

Answer ✓ Key

First Worksheet 2

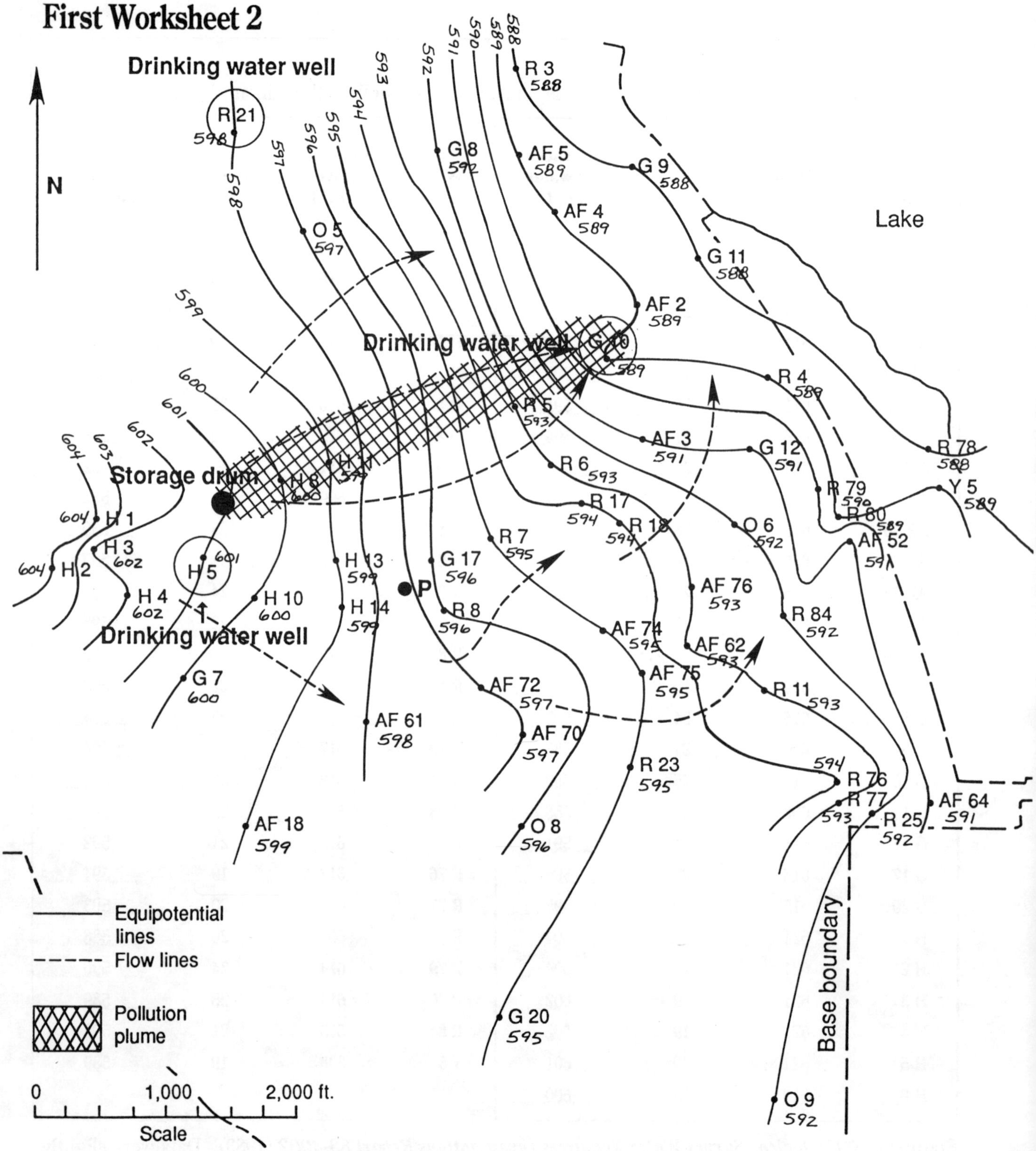

Figure 2. Map of military base showing locations of selected wells; adapted from *U.S. Geological Survey Water Resources Investigations Report 83-4002* (1983).

Answer Key

Second Worksheet 2

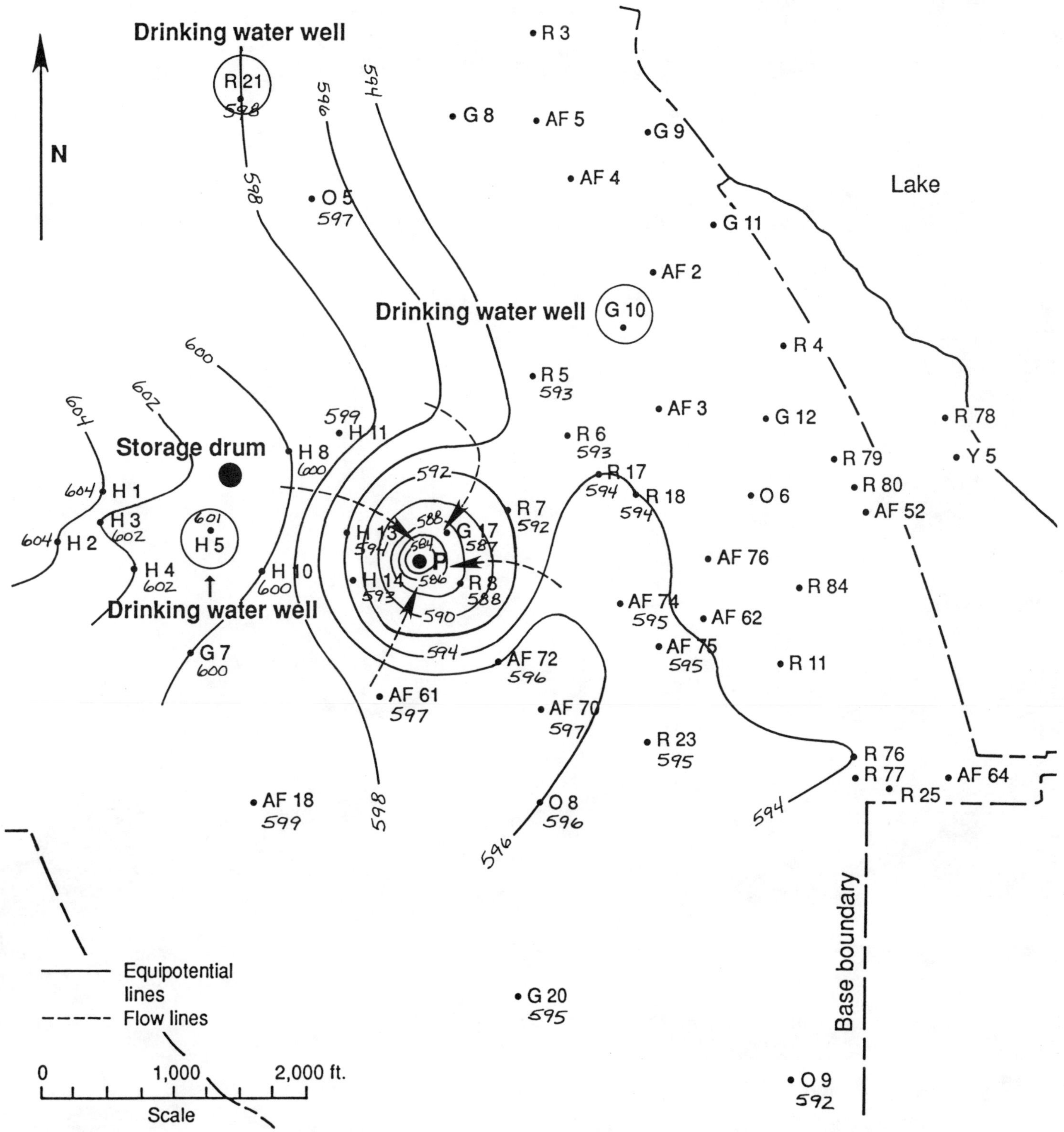

Figure 2. Map of military base showing locations of selected wells; adapted from *U.S. Geological Survey Water Resources Investigations Report 83-4002* (1983).

World Time-Day Calculator

JOSEPH GRANDE
Abington Junior High School
Abington, Pennsylvania

Objective

To calculate the time and day in selected major cities in the Northern Hemisphere by constructing and using a world time-day calculator.

Time required

Two class periods

Materials

per student
Compass (for drawing circles)
Protractor
Metric ruler
Scissors
Brass clip or other similar fastener
2 pieces of light-colored construction paper: different colors; one should be 20 cm x 25 cm (8 in. x 10 in.), and one 18 cm x 18 cm (7 in. x 7 in.)
Map or globe of the world (optional)

Background

Time zones around the Earth are bounded by imaginary lines called lines of **longitude**, or meridians. On a globe or map of the world, they are the lines that are perpendicular to the equator. They are farthest apart at the equator and gradually come closer together, finally meeting at the North and South Poles. Meridians are numbered in degrees east and west from the **prime meridian**, which is 0° longitude, and which passes through Greenwich, England.

Meridians divide the Earth's surface into 24 time zones. In general, each time zone is 15° of longitude wide and represents one hour of time. The lines that bound the time zones sometimes show indentations to allow for geographic regions. Time zones can be numbered, starting at the prime meridian (0° longitude), which extends through the center of time zone 1. Halfway around the Earth, 180° from the prime meridian, is the **international date line** (IDL). Points on the west longitude side of the IDL are one day earlier than points on the east longitude side. Because the IDL should not pass through populated areas, it zigzags to avoid countries and islands.

Procedure

1. Using the larger piece of construction paper, cut out a circle with an 8.5-cm (3 1/4-in.) radius. This is circle A.

2. Using the other piece of construction paper, cut out another circle with a 7-cm (2 3/4-in.) radius. This is circle B.

3. On circle A, draw a diameter. Place a protractor on the line and mark off 15° intervals. Draw diameters through these points. See Figure 1.

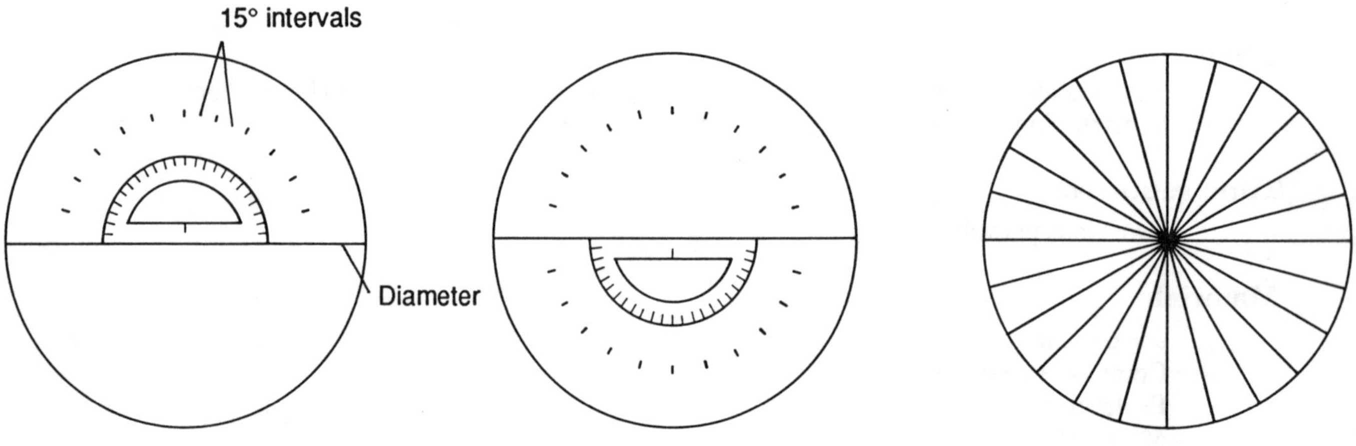

Figure 1. Follow these steps to make circle A.

4. On circle B, draw a circle with a radius of 6 cm (2 1/4 in.). Repeat step 3, marking off 15° intervals on circle B, but when you draw the lines, end them on the 6-cm radius circle. See Figure 2.

5. Fasten circle B on top of circle A with a brass fastener through their centers. Circle B represents the Northern Hemisphere, the head of the fastener is the North Pole, and the 6-cm circle is the equator. The 15° interval lines on circle B divide the circle into 24 time zones. Circle A is the hour circle.

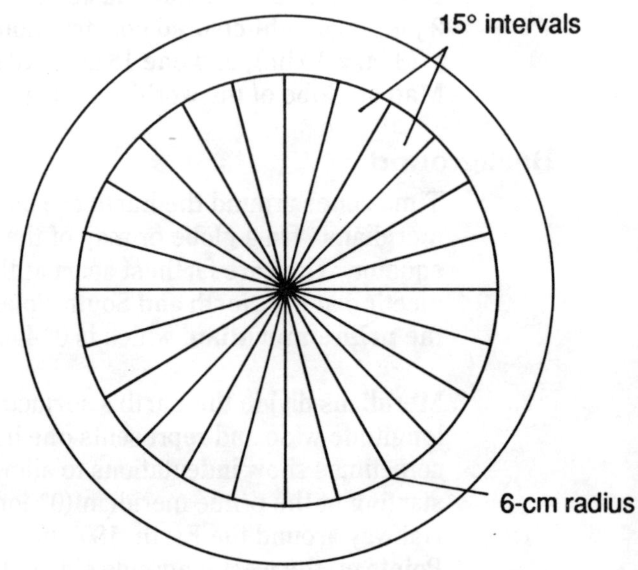

Figure 2. Circle B.

6. Mark circle B as shown in Figure 3. Parts of the circle are labeled in Figure 3 to help you understand the markings, but the labels do not need to be marked on your circle. Write the longitude markings. The prime meridian is located at the center of the zone that starts at 0° longitude. Each 15 degrees of longitude is therefore marked at the center of each zone. The meridians are drawn 7.5° on either side of the degree mark. Draw a darker line to represent the international date line at 180° to indicate the *place* the day changes. Mark the locations of the four cities on the circle. Number the 24 time zones. Draw a counterclockwise arrow to show the direction the Northern Hemisphere is rotating.

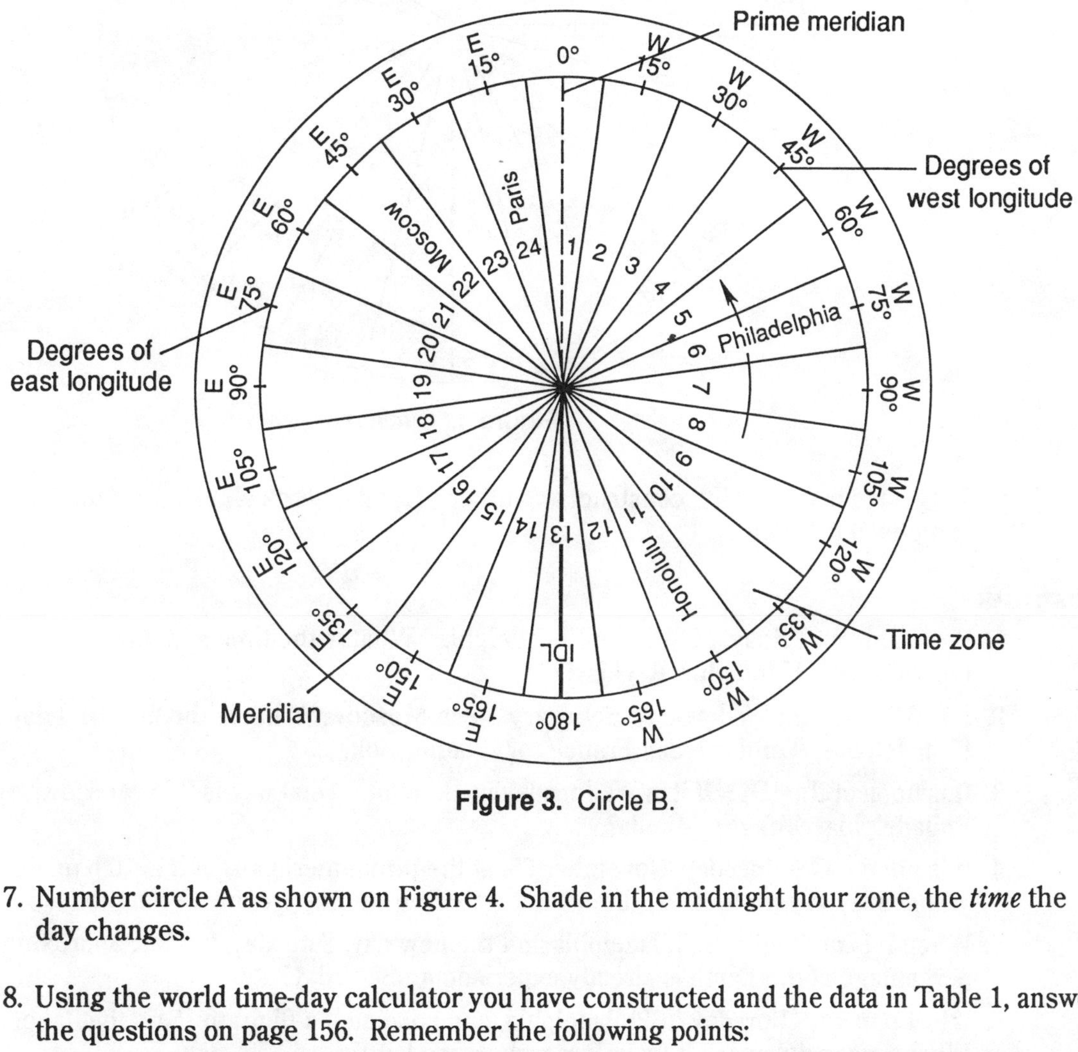

Figure 3. Circle B.

7. Number circle A as shown on Figure 4. Shade in the midnight hour zone, the *time* the day changes.

8. Using the world time-day calculator you have constructed and the data in Table 1, answer the questions on page 156. Remember the following points:

   The new day begins on the east longitude side of the IDL and exists clockwise around the world to the midnight hour.

   When using the calculator, always set the time zone directly *in line* with an hour zone, disregarding any minutes involved in the calculation. The minutes can be added on to any other cities' hour zones.

   During daylight savings time (from April to October), subtract one hour when calculating time between a U.S. city and a non-U.S. city.

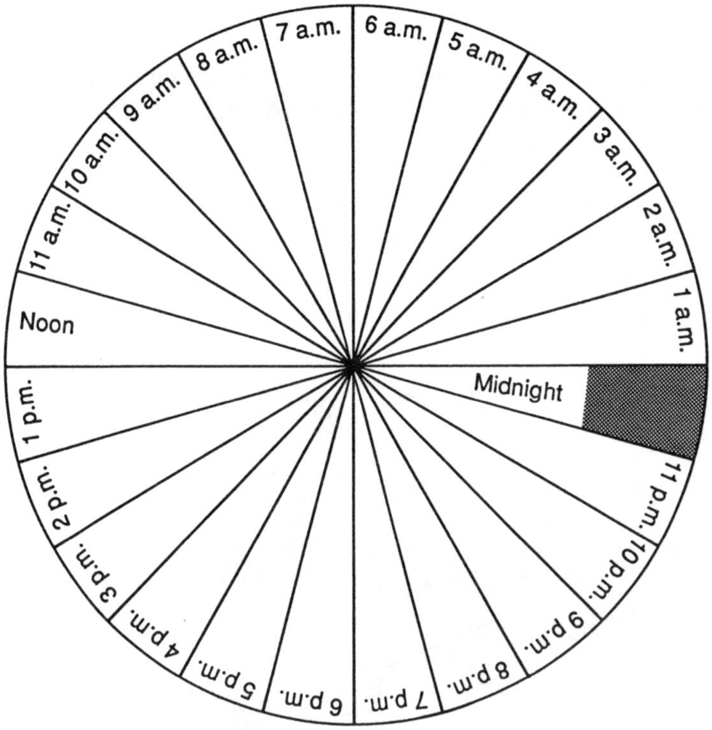

Figure 4. Circle A.

9. Suggested extra credit: construct a Southern Hemisphere calculator on the back of the completed calculator.

Questions

1. It is 4 a.m. on Friday, May 12, in Philadelphia. What is the time and day in Paris? Peking? Los Angeles? Honolulu? Reykjavik?
2. It is 6:22 p.m. on Wednesday, February 20, in Moscow. What is the time and day in Philadelphia? London? San Francisco? Vladivostok?
3. It is noon at the IDL. If it is Wednesday in Honolulu, what day is it in Moscow? Philadelphia? Tokyo? Manila?
4. It is noon on Wednesday, November 25, at the prime meridian. If it is 10 p.m. on Wednesday at your location, then what would be your approximate longitude?
5. When it is midnight in Philadelphia and the new day, Saturday, begins, approximately what percentage of the Earth is already experiencing Saturday?
6. It is 4 p.m. on Thursday in Philadelphia, and you catch a flight to Paris that takes six hours. What time and day will it be in Paris when you land?

Table 1
Time zones of major cities in the Northern Hemisphere.

Zone	City	Zone	City	Zone	City
1	Accra, Ghana	9	Los Angeles, Calif., U.S.	18	Saigon, Vietnam
10	Anchorage, Alaska, U.S.	24	Madrid, Spain	9	San Francisco, Calif., U.S.
23	Athens, Greece	17	Manila, Philippines	9	Seattle, Wash., U.S.
24	Berlin, Germany	7	Mexico City, Mexico	16	Seoul, South Korea
23	Cairo, Egypt	6	Miami, Fla., U.S.	17	Shanghai, China
7	Chicago, Ill., U.S.	6	Montreal, Quebec, Can.	24	Stockholm, Sweden
8	Denver, Colo., U.S.	22	Moscow, U.S.S.R.	19	Tashkent, U.S.S.R.
8	Edmonton, Alberta, Can.	7	New Orleans, La., U.S.	23	Tel Aviv-Jaffa, Israel
5	Halifax, Nova Scotia, Can.	6	New York, N.Y., U.S.	5	Thule, Greenland
5	Hamilton, Bermuda	10	Nome, Alaska, U.S.	16	Tokyo, Japan
6	Havana, Cuba	24	Paris, France	6	Washington, D.C., U.S.
23	Helsinki, Finland	17	Peking, China	7	Winnipeg, Manitoba, Can.
17	Hong Kong, China	6	Philadelphia, Pa., U.S.	9	Vancouver, B.C., Can.
11	Honolulu, Hawaii, U.S.	1	Reykjavik, Iceland	15	Vladivostok, U.S.S.R.
1	London, England	24	Rome, Italy		

Notes: Time-zone numbers for the cities were derived from information in the *Hammond World Atlas* (Maplewood, N.J.: Hammond, 1988). The standard time followed in some countries does not always coincide with zone time.

Answer Key

Questions

1. Paris — 9 a.m. Friday (remember daylight savings time)
   Peking — 4 p.m. Friday
   Los Angeles — 1 a.m. Friday
   Honolulu — 11 p.m. Thursday
   Reykjavik — 8 a.m. Friday

2. Philadelphia — 10:22 a.m. Wednesday
   London — 3:22 p.m. Wednesday
   San Francisco — 7:22 a.m. Wednesday
   Vladivostok — 1:22 a.m. Thursday

3. Moscow — Thursday
   Philadelphia — Wednesday
   Tokyo — Thursday
   Manila — Thursday

4. Your approximate longitude would be 142.5° E to 157.5° E.

5. Approximately 70% of the Earth is experiencing Saturday.

6. It will be 4 a.m. on Friday in Paris.

Determining the Earth's Circumference

GENE HINMAN
Cornell College
Mt. Vernon, Iowa

Objective

To determine the approximate circumference of the Earth using geometric principles and mathematical proportions.

Time required

Two class periods
This lab should be conducted at noon on either the autumnal equinox, September 22 or 23, or the vernal equinox, March 20 or 21. At these times, the sun is directly over the equator, the position needed so that students can use the sun's rays to make angular measurements.

Materials

per student
Notebook paper: 8 1/2 in. x 11 in.
2 pencils
Notebook or other firm, flat surface on which to lay the notebook paper
Compass (for drawing circles)
Metric ruler
String: 38 cm (15 in.) long

Background

Note: Students will need pre-lab instruction in locating north at noon, understanding ratios and proportions, measuring angular distance, and understanding the relationship between angles formed by a transverse line that cuts across parallel lines.

Eratosthenes (276-196 B.C.), a Greek mathematician and astronomer, was able to measure the Earth's circumference, the distance around the Earth, by applying the following assumptions, some simple observations, and mathematics.

He assumed that the Earth was a sphere, an idea that was generally believed during his time, and that the sun was larger than the Earth. He also assumed that the sun was so far away that its rays were parallel when they reached the Earth.

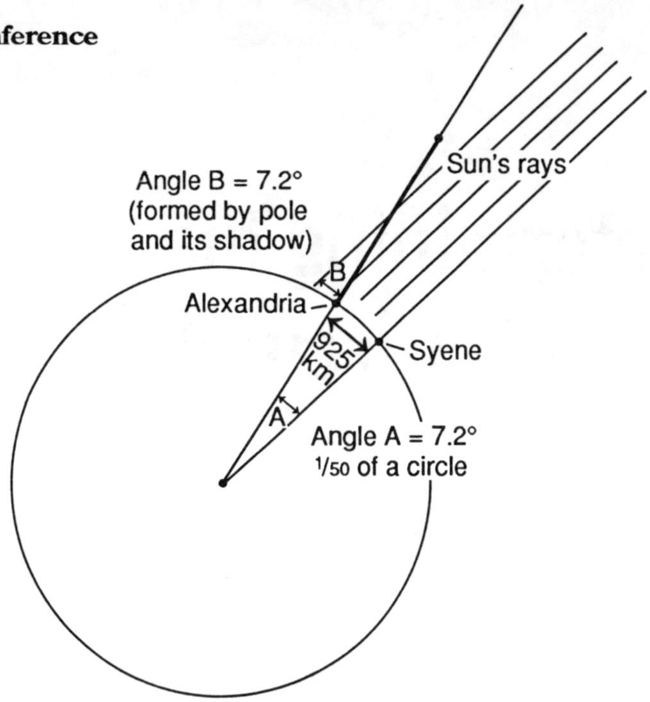

Figure 1. Eratosthenes' method. Angle A = Angle B. Angle A = 7.2° or 1/50 of a circle (360°). The distance between Alexandria and Syene is 925 km: 925 km x 50 = 46,250 km, which is the approximate circumference of the Earth.

He observed that on the summer solstice, June 21 or 22, the noon sun was directly overhead in Syene, Egypt, but lower on the horizon in Alexandria, a town directly north of Syene.

He knew from geometry that a measurement of the angle of the sun's rays at Alexandria during the solstice would be equal in size to the angle at the Earth's center between Alexandria and Syene (Figure 1). He reasoned that this angular measurement was a fraction of the circumference of a full circle (360°), and the linear distance between the two towns, the arc of this circle segment, would be in the same proportion to the Earth's circumference, the distance around the Earth.

Because schools are rarely in session during the summer solstice, Eratosthenes' experiment has been modified to work during either the autumnal equinox, September 22 or 23, or the vernal equinox, March 20 or 21.

Procedure

The teacher or students should determine the distance in kilometers from the equator to the location of the investigation. A globe or medium-scale map will provide the information, which is needed for Part III.

Part I. Using Eratosthenes' method (Figure 2)

1. Using a compass, draw a circle 10 cm (4 in.) in diameter. This circle represents a cross section of the Earth.

2. Draw an equator through this cross section and extend it 5 cm (2 in.) beyond the circle. The extension is the "path" that a ray of sun follows from the sun to the equator.

3. Pick a point on the circle some distance from the equator, and draw a line from that point to the center of the circle. Extend the line 5 cm (2 in.) beyond the point on the circle. The point represents a place some distance from the equator, and the line represents a perpendicular from that point to the center of the Earth.

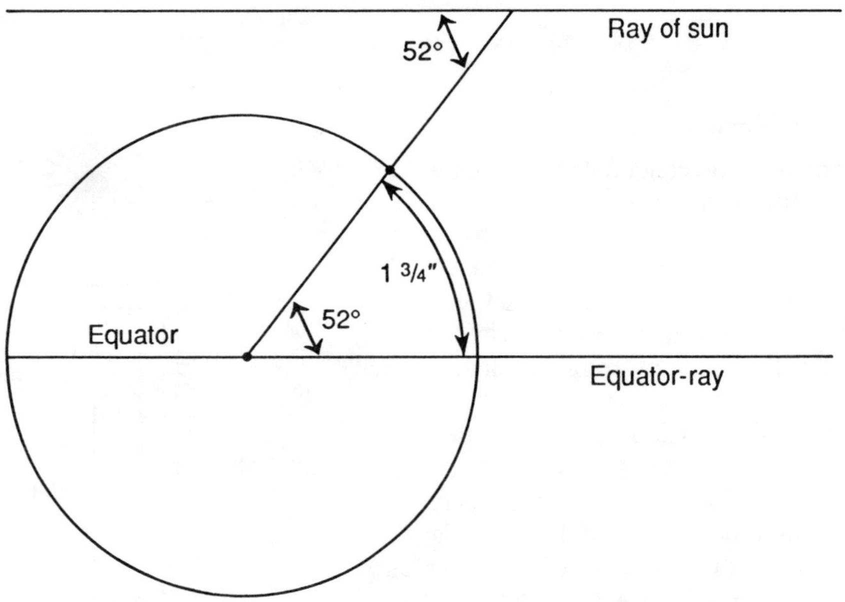

Figure 2. Sample calculation using Eratosthenes' method.

4. Using a piece of string, measure the distance from the point to the equator along the circumference of the circle. Lay the string on the circle between the point and the equator, then place it next to the ruler to determine the actual length.

5. Record the measurement: ....

6. Measure the angle at the center of the circle made by the intersection of the equator and the line drawn perpendicular from the distant point on the circle.

7. Record the measurement: ....
   Of course Eratosthenes could not measure this angle, but he could determine what it was by measuring an equal angle at the Earth's surface.

8. Draw another ray of sun that is parallel to the equator-ray so that the second ray intersects the extension of your line projecting from the center of the circle.

9. Measure the angle where these two lines intersect.

10. Record the measurement: ....

11. Compare the measurement from step 7 with that from step 10. They should be the same.

12. To calculate the circumference of your Earth, you must enter your data in an equation that represents the relationship of your measured data to that of your Earth.

$$\frac{\text{distance from equator to point}}{\text{circumference of your Earth}} = \frac{\text{angular distance between equator and point}}{\text{angular measure of a circle}}$$

$$\frac{\text{measurement in step 5}}{\text{(unknown distance)}} = \frac{\text{angle in step 10}}{360°}$$

Check your answer by measuring the circle with the string.

Part II. Collecting data

Data must be collected at noon on either the autumnal or the vernal equinox.

1. Place the notebook paper on a notebook, and lay both on the ground so that they are oriented with the length roughly north-south.

2. Hold a pencil perpendicular to the paper at either southern corner, and rotate the notebook (and paper) until the shadow of the pencil falls along the long edge of the paper, as shown in Figure 3. Look down on the top of the pencil to make sure it is vertical.

3. With a second pencil, mark both ends of the shadow, and take the paper into the lab. An accurate plotting of the distance between both ends is essential.

4. Draw a line connecting the top and bottom of the pencil shadow, as shown in Figure 4. This will be called the shadow line.

5. At either end of the shadow line, draw a perpendicular line the same length as the pencil used to make the shadow. This will be called the pencil line.

6. Draw a line connecting the opposite ends of both lines. It will be the hypotenuse of the resulting triangle.

7. Measure the angle made by the intersection of the hypotenuse and the pencil line. This angle should equal the local latitude and will be called the sun angle.

8. Record the angle: ....

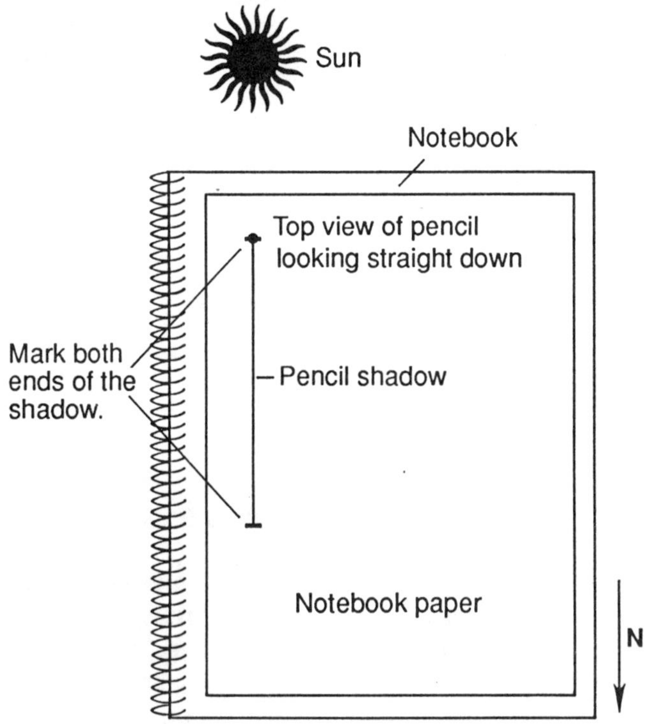

Figure 3. Setup for collecting data outside.

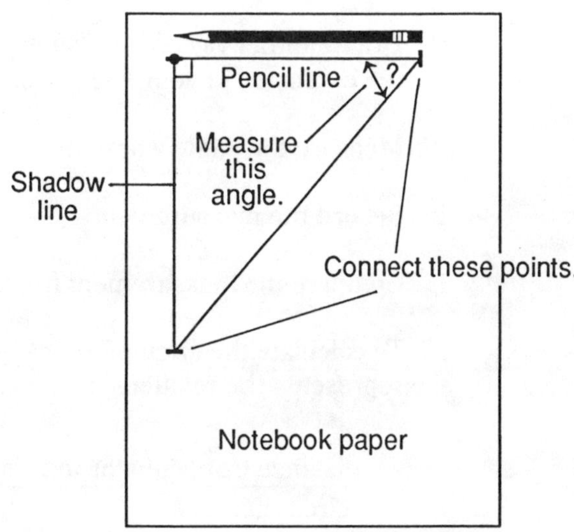

Figure 4. Setup for analyzing data inside.

Part III. Analyzing data

Your data collection was to determine the angular measure between your location and the equator. This angle is the latitude of your location.

To complete the calculations for the Earth's circumference, you must know the distance in kilometers from your location to the equator.

Enter these data in the equation below to calculate the Earth's circumference:

$$\frac{\text{distance to equator}}{\text{circumference of Earth}} = \frac{\text{sun angle}}{360^\circ}$$

Questions

1. Why is it essential that the measurements be taken at noon on one of the equinoxes?
2. Why is it essential that the sun is a great distance from the Earth?
3. Why are the two angles measured in Part I the same?
4. What is the latitude of your school?

Answer Key

Questions

1. These are the only times when the sun is directly over the equator.

2. Because of the sun's great distance from the Earth, the rays of sunlight reaching Earth may be considered parallel.

3. When two parallel lines are intersected by a third line, the alternate interior angles are equal.

4. The latitude is the angular measure recorded in Part II, step 8.

Observing Our Solar System

DENNIS W. SUNAL

The University of Alabama
Tuscaloosa, Alabama

Objectives

To construct a two-dimensional model of our solar system from measurements made in the current sky.

To construct and/or use angle-measuring devices.

To observe and record angle measurements from the sun to visible planets and constellations in the current sky.

To relate the classic solar system model to the planets and constellations visible in the current sky.

Time required

Two class periods plus a planetarium visit or nighttime sky observation

Materials

per group of 2 students

Astrolabe: Commercial versions come in different designs, some with straws attached to protractors. To construct an astrolabe, you will need the following materials:

- Heavy cardboard or posterboard: 40 cm x 40 cm (16 in. x 16 in.)
- Protractor
- Brass clip or other fastener
- String: 30 cm (12 in.) long
- Large washer or other similar weight
- Scissors
- Metric ruler

per student

Monthly star map found in astronomy magazines or science teaching journals (*Astronomy, Science and Children*, or *Sky and Telescope*)

Protractor

Background

Note: Students should have an understanding of planes and angles.

A common way to represent the solar system from earliest times has been to draw concentric circles and put the planets somewhere on them as dots or small circles. After seeing these enough times, one might begin to wonder, "How do they know?" Textbooks provide general images of the solar system, but they do not show what it looks like today. In this investigation, you will illustrate the solar system to show what it looked like on the date of your observations.

Procedure

Part I. Measuring angles

Section A. Constructing an astrolabe (optional for classrooms with no commercial astrolabes)

Basic astronomical measurements include length, time, angle, mass, and properties of light. For this part of the investigation, you will use only angle measurement. The most practical device for measuring angles is a protractor. Devices incorporating a protractor for measuring angles between objects in the sky are called sextants. In ancient times, instruments called astrolabes were used. To construct one, follow the steps below (Figure 1).

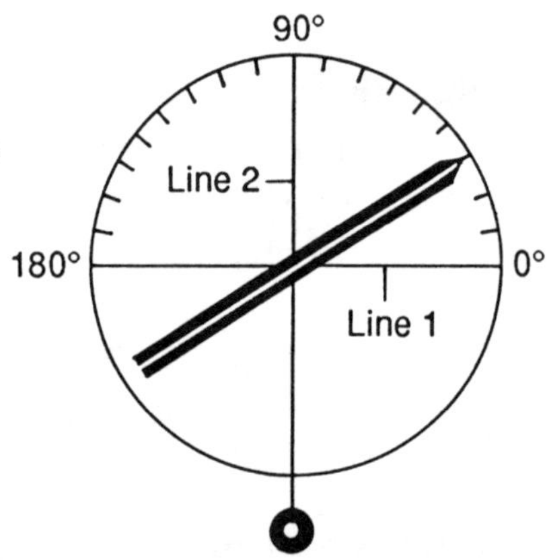

Figure 1. Astrolabe.

1. Cut a piece of posterboard or cardboard about 40 cm (16 in.) in diameter.
2. Draw two perpendicular lines on the circle.
3. Mark graduations from 0° to 180° at 10-degree intervals.
4. Construct a pointer 40 cm x 5 cm (16 in. x 2 in.). It should be a slightly shorter length than the diameter of the circle, as shown in Figure 1. Draw a line down the middle of the pointer.
5. Attach the pointer to the center of the circle with a brass clip or other fastener.
6. Attach a weight to one end of a string, and attach the other end of the string to the brass clip at the center of the cardboard.

Section B. Using an astrolabe

Astronomers used an astrolabe in two different ways. One way is to measure an object's altitude or height above the horizon (Figure 2). When the hanging string is parallel to line 2, line 1 is parallel to the horizon. Thus, sighting along the line on the pointer will produce a reading of an object in degrees above the horizon (altitude).

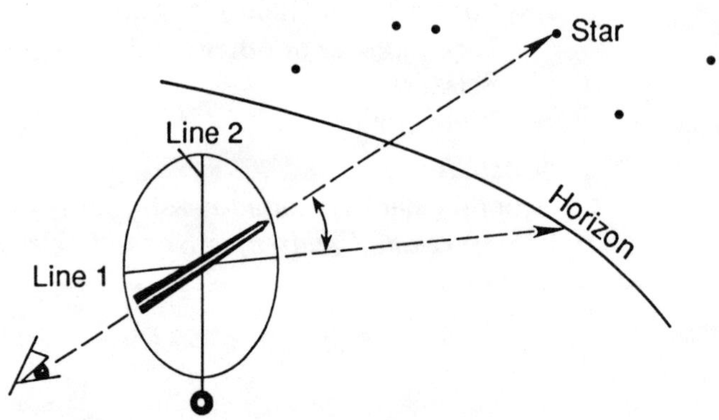

Figure 2. Measuring altitude using a student-constructed astrolabe.

A second way is to measure the angular separation of two objects (Figure 3). The string and weight are not used in this type of measurement. Point line 1 at object A and the pointer at object B. The angle read at the end of the pointer is the angular separation of A and B.

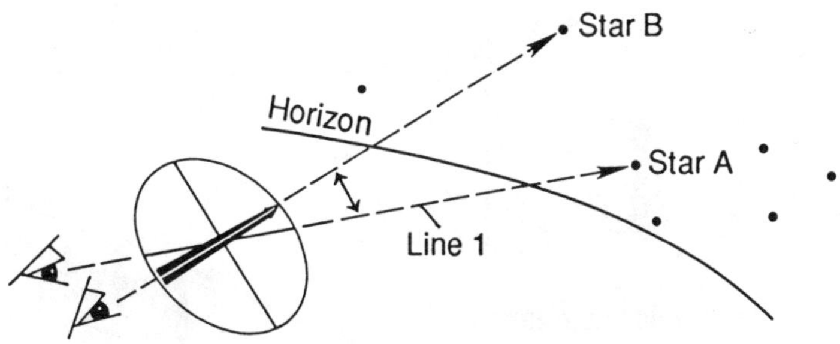

Figure 3. Measuring the angular separation of two objects using a student-constructed astrolabe.

Since the sun is the central object in the solar system, you will make all measurements in relation to the sun. After sunset, you can estimate where the sun is below the horizon and measure the angle between the sun and any sky object.

For greater accuracy, the moon or a bright planet, which can be seen with the sun in the daytime sky, can be used as an intermediate reference object in making your measurements (Figure 4).

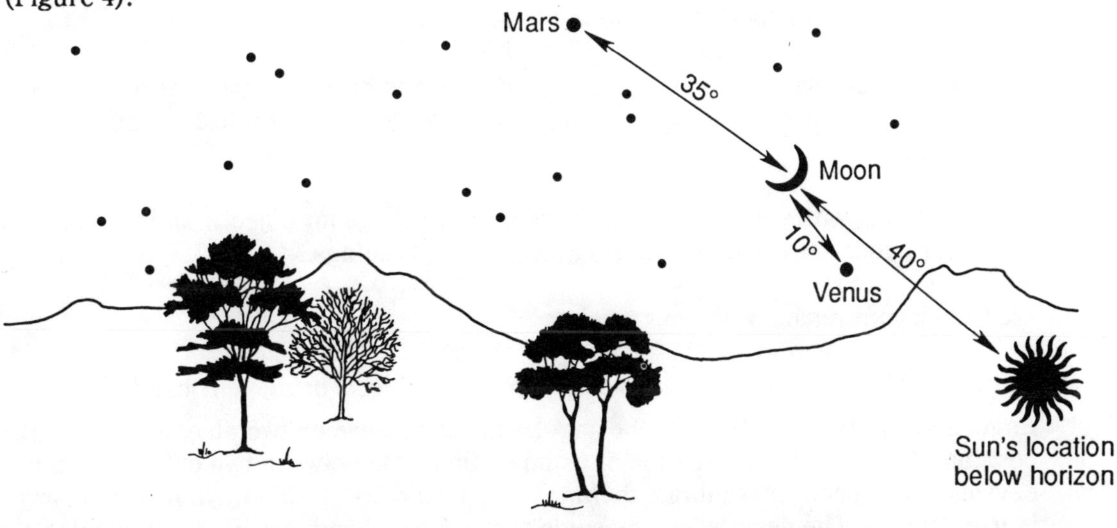

Figure 4. Measurement using reference objects.

Sample measurements:
sun to moon = 40° (measured before sunset) **CAUTION:** *Do not look directly at the sun.*
moon to Mars = 35° (measured after sunset)
moon to Venus = 10° (measured after sunset)

Sample sun angle calculations:
sun to Venus = sun to moon angle (40°) - moon to Venus angle (10°) = 30°
sun to Mars = sun to moon angle (40°) + moon to Mars angle (35°) = 75°

Section C. Measuring altitude using an astrolabe

Altitude is defined as the angular measurement of an object above the horizon. Using a commercial astrolabe or a student-constructed astrolabe, determine the altitude of an object above a table top (Figure 5). ***Students will need partners.***

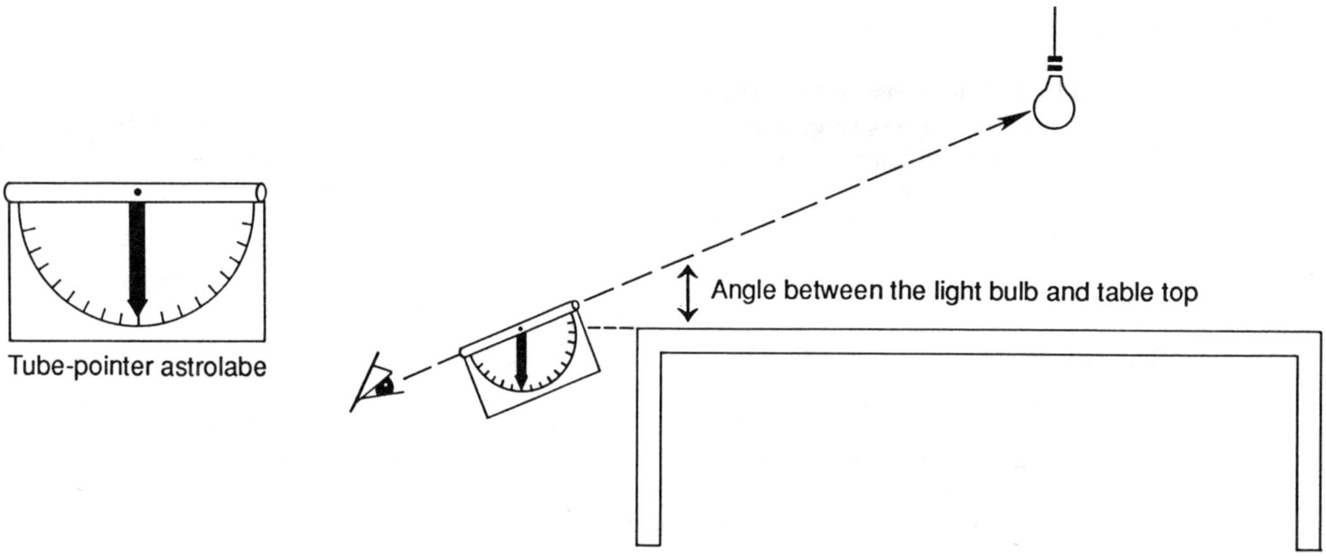

Figure 5. Measuring altitude with a commercial tube-pointer astrolabe.

1. Sight along the table surface through the tube or astrolabe so that your eye is level with the table. The astrolabe should read 0°.

2. Point the astrolabe at an object above the table and have your partner read the astrolabe scale angle. Record the angle for object 1 on Worksheet 1, Table 1, on the line beside "First reading."

3. Alternating with your partner, take two more readings for object 1 and three for object 2. Record your data in the table, and average your readings.

4. Check your results with your teacher.

Section D. Measuring the angular separation of two objects using your hand

Angular separation is defined as the angular distance between two objects from an observer's position. You can use an astrolabe or estimate the angle between two objects. You have several devices handy once you calibrate them to estimate angles. Hold out your hand at arm's length in front of you. The devices include some parts of your hand: two of your knuckles, thumb, fist, and the span of your hand (from thumb-tip to the tip of your little finger).

As a first approximation, your fist is about 8°. Since the length of people's arms and the size of their hands vary, however, you will need to figure out your own hand calibrations.

Use either of the following methods to calibrate parts of your hand. *Students will need partners.*

Method 1. Calibration by astrolabe

1. Hold the specific part of your hand that you want to calibrate at arm's length in front of you. Direct your partner to put marks on a blackboard or to tape thin paper strips on a wall so that the space between the marks equals the width of your hand part as you look at your hand in relation to the blackboard or wall. Stay in the same position and measure the angular distance between the marks with your astrolabe. Repeat with other hand parts. Record your angular measurements on Worksheet 1, Table 2. Write your angles under "Method 1."

2. Draw a diagram of your procedure.

Method 2. Calibration using a simple geometric relationship

An object viewed from a distance that is 57.4 times its diameter covers an angle of 1°. Thus, two pencils separated by one cm and viewed from 57.4 cm (23 in.) away, have an angular separation of 1°.

1. Set a metric ruler on its edge 57.4 cm away from your eye (114.8 cm for a 2-cm [1°] separation). Hold the specific part of your hand that you want to calibrate at arm's length in front of you. Determine the values to put on Worksheet 1, Table 2. Write your angles under "Method 2."

2. Draw a diagram of your procedure.

After calibrating the parts of your hand, use one of the parts to determine the angular separation of two sets of objects. Record your angles on Worksheet 1, Table 3.

Section E. (Optional) Measuring the altitude of a light source by construction

In addition to using an astrolabe, you can measure altitudes of light sources using shadows. To do so, you use an object such as a short stick (pencil), peg, or nail. The object, which is called a gnomon, casts a shadow, and you compare the length of the shadow with the length of the gnomon. Then by drawing a scale model, the altitude of the sun (or light source) can be determined (Figure 6).

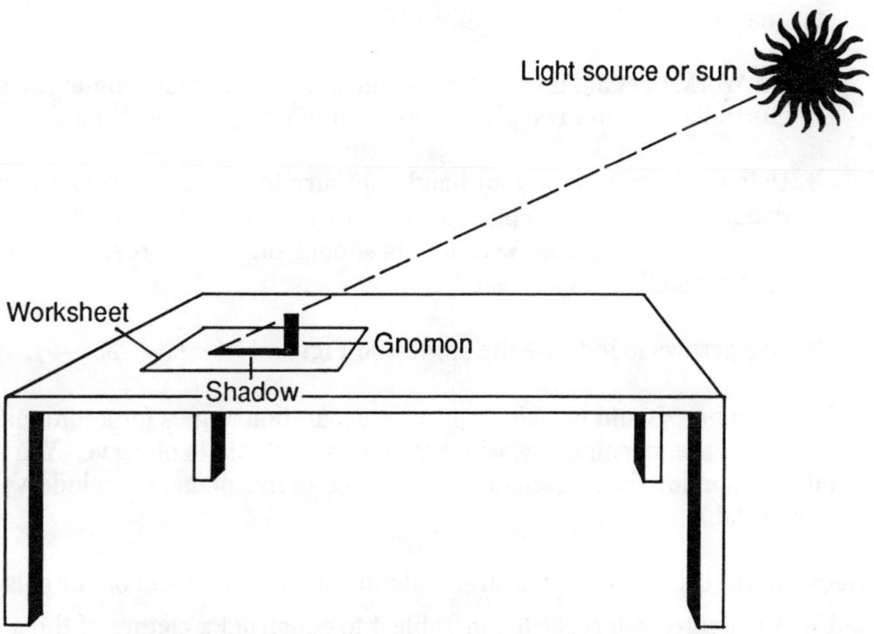

Figure 6. Measurement using a gnomon.

1. Set Worksheet 2 on the table, and set a gnomon upright at the location marked on the worksheet. See Figure 7.

2. Position the paper so that the gnomon's shadow falls on the area indicated on the worksheet. Mark the length of the shadow on the worksheet.

3. Lay the gnomon down on the sheet at right angles to the shadow line, and mark off the length of the gnomon.

4. Complete the triangle on the worksheet by drawing the remaining side, connecting the tip of the gnomon with the end of the shadow.

5. Measure the angle opposite the gnomon, using a protractor, and record your results on the worksheet. This is a safe and sure measurement of the altitude accomplished by geometric construction.

6. Check your results using an astrolabe; record the angle on the worksheet.

7. Check your results using your thumb, knuckles, or fist; record the angle on the worksheet.

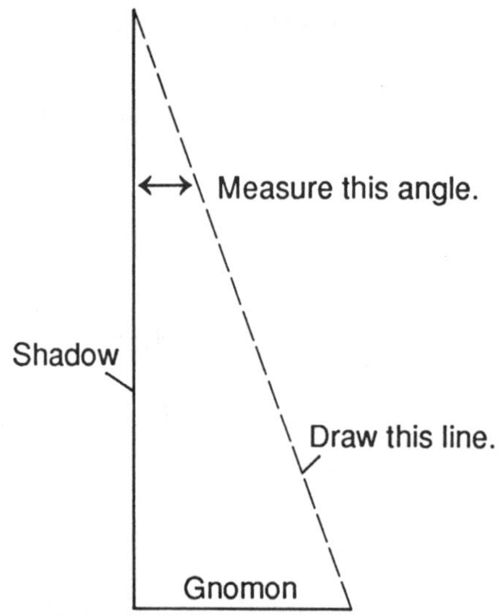

Figure 7. How to use a gnomon.

Part II. Constructing a view of our solar system from the Earth

Section A. Measuring angles between objects in the current sky

For this part of the investigation, you will need to observe the evening sky just after sunset or a planetarium that is set for early evening. Before beginning, you may want to look at a monthly star map to help identify a few objects.

1. On Worksheet 3, Figure 11, draw the approximate location of the sun and at least two constellations and two planets, as seen in tonight's southern sky.

2. Using an astrolabe or your hand, measure the angle between the sun and at least two constellations and two planets. Record your angles on Worksheet 3, Table 4. If needed, review Part I, Section A. Students should share their results with the class so that up to four constellations are listed.

3. Use arrows to indicate the angles on Figure 11 (e.g., ←38°→).

Your teacher should be able to provide separation angles for additional constellations and planets in the morning sky, which you were not able to observe. You should have complete information for six constellations and three to five planets. Include Venus, Jupiter, and Saturn in your table.

Section B. Using angle measurements to construct a model of our solar system

Use the angles you recorded in Table 4 to construct a picture of the solar system on Worksheet 4, Figure 12.

1. On Worksheet 4, write the names of the constellations on the constellation semicircle at the separation angle you observed them from the sun, as seen from the Earth.

2. Put the planets (use dots) on their orbital path at the angle you observed them from the sun. When you finish, this is what the solar system looks like at this moment. If measurements are made at a future time, what changes would you expect?

Questions

Answer the following questions, using your model of the solar system. Work in small groups and discuss them with the class.

1. What is the date the solar system looked like your model?
2. In what directions would a spacecraft find no planets when traveling out from the Earth? Give angles from the sun.
3. A spacecraft traveling out from the Earth at 90° away from the sun, in the evening sky, will come closest to which planet? Toward what constellation will the spacecraft be traveling?
4. Observation has found that all planets revolve in the same direction around the sun, counterclockwise. The speed of revolution decreases the farther a planet is from the sun. Superior planets are those with orbits outside the orbit of the Earth, and inferior planets are those with orbits inside the orbit of the Earth. When the Earth overtakes a superior planet, the planet is either in conjunction with the sun — when viewed from Earth, it is aligned with the sun — or in opposition to the sun — it is on the opposite side of the Earth from the side the sun is on, exactly 180° from the sun. When the Earth is overtaken by an inferior planet, the planet is said to be in conjunction with the sun (Figure 8).

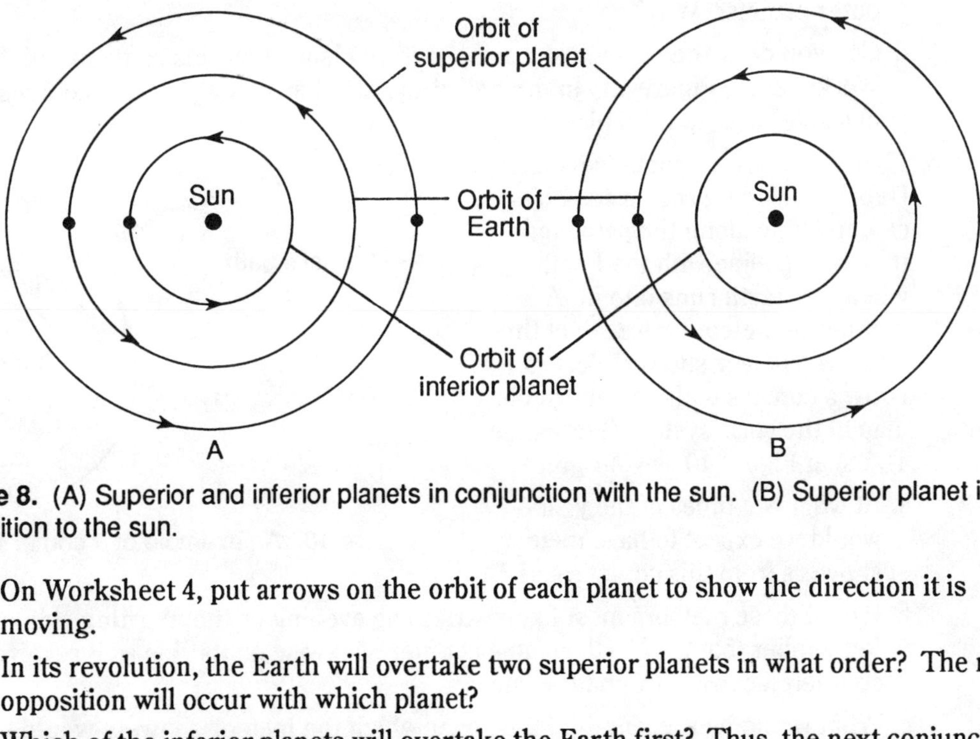

Figure 8. (A) Superior and inferior planets in conjunction with the sun. (B) Superior planet in opposition to the sun.

   a. On Worksheet 4, put arrows on the orbit of each planet to show the direction it is moving.

   b. In its revolution, the Earth will overtake two superior planets in what order? The next opposition will occur with which planet?

   c. Which of the inferior planets will overtake the Earth first? Thus, the next conjunction will occur with which planet?

5. a. If you landed in a spacecraft on one of the planets in your model and looked at the sky on the night you made your astrolabe measurements, would you find the planets in the same constellation, as seen from the Earth?

   b. Using your model map of the solar system (Worksheet 4), pick two planets on which to land your spacecraft. Write their names on Worksheet 5, Table 5, beside "Planet 1" and "Planet 2." Pretend you have landed on planet 1 and are observing objects in the sky.

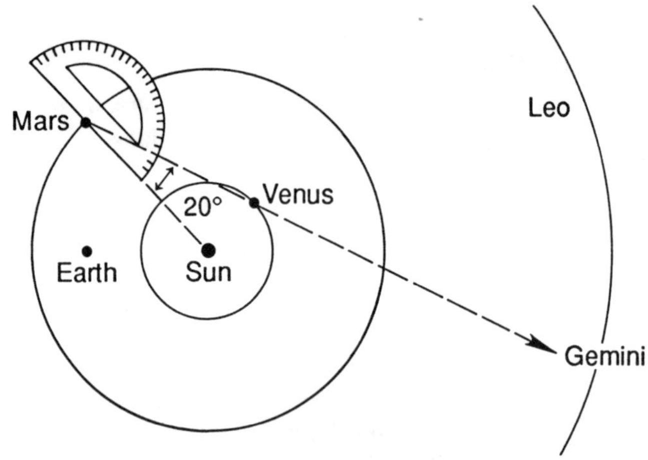

Figure 9. Measuring the angle between the sun and Venus from your location on Mars.

The objects are listed in the first column in Table 5. Using a protractor on Worksheet 4, measure the angular separation between each object and the sun. Record your angles in Table 5. Repeat for planet 2.

For example, you have landed on Mars (planet 1), and you are to measure the angular separation of Venus (object 2 in the first column) and the sun. You would measure the angle between Venus and the sun with the protractor centered on Mars. Figure 9 shows 20° separation. Venus will be seen in Gemini from Mars but in Leo from Earth.

c. Patterns are found when comparing these measurements and those from the Earth. What is noted about the positions of the Earth and Venus when measuring from the outer planets? Why?

d. Can you draw the evening sky as seen from planet 1 on this same night? Try it on Worksheet 5, Figure 13. In the half circle, draw as many planets and constellations as possible, as seen from planet 1.

6. Figure 10 shows a comet's path. Debris, including rock-sized objects, can be found along the path, and they may collide with the Earth when the Earth runs into it. A large number of meteors are seen at this time. A "meteor shower" occurs. Draw a comet's path on your model map of the solar system (Worksheet 4), using Figure 10 as your guide.

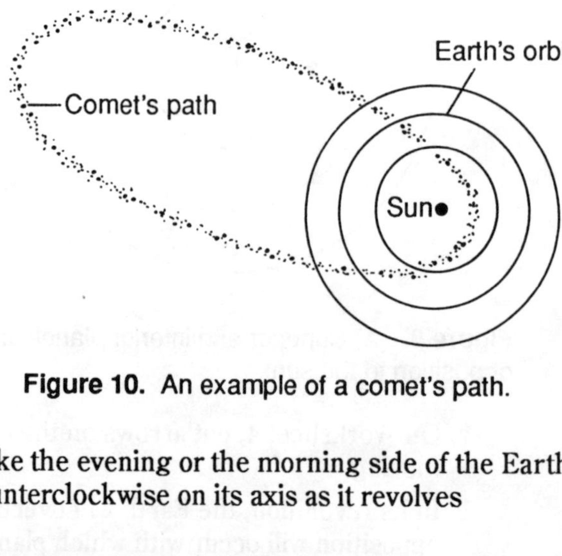

Figure 10. An example of a comet's path.

a. At what two times of the year would we expect to have meteor showers from this comet's path?

b. Would these meteors most likely strike the evening or the morning side of the Earth? Remember that the Earth rotates counterclockwise on its axis as it revolves counterclockwise around the sun.

c. What would be the length of time over which the meteor showers would be seen — days, months, or years? Why would the number of meteors vary from year to year?

Additional resources

Abrams Planetarium, Michigan State University, East Lansing, Mich. *Sky calendar.*

Kalmbach Publishing Co., Waukesha, Wis. *Astronomy Magazine* and *Odyssey.*

National ScienceTeachers Association, Washington, D.C. *Science and Children, Science Scope,* and *The Science Teacher.*

Sky Publishing Co., Cambridge, Mass. *Sky and Telescope.*

Name ....

Class ....

Date ....

Worksheet 1

Table 1
Measuring altitude using an astrolabe.

	Altitude of object 1	Altitude of object 2
First reading	....	....
Second reading	....	....
Third reading	....	....
Average reading	....	....

Table 2
Calibrating parts of your hand.

	Method 1	Method 2
Two of your knuckles	....	....
Your thumb width	....	....
Your fist	....	....
Span of your open hand (from thumb-tip to the tip of your little finger)	....	....

Table 3
Measuring the angular separation of two objects using parts of your hand.

	Set 1		Set 2	
Objects (name two per set)	....	....	....	....
Angular separation of the two objects	....		....	

Observing Our Solar System

Worksheet 2

Name ....

Class ....

Date ....

		Angle
Altitude (by construction)	=	....
Altitude (by astrolabe)	=	....
Altitude (by hand)	=	....

Mark shadow along here.

• Place gnomon upright here.

Lay gnomon along here.

Name ....

Class ....

Date ....

Worksheet 3

Overhead Date ....

East South West

Figure 11. Tonight's southern sky, from the inside looking out.

Table 4
Measuring angles between objects in the current sky.

Constellation	Angle from sun to center of constellation	Planet	Angle from sun to planet
....	....	....	....
....	....	....	....
....	....	....	....
....	....	....	....
....	....	....	....
....	....		

Worksheet 4

Name ....

Class ....

Date ....

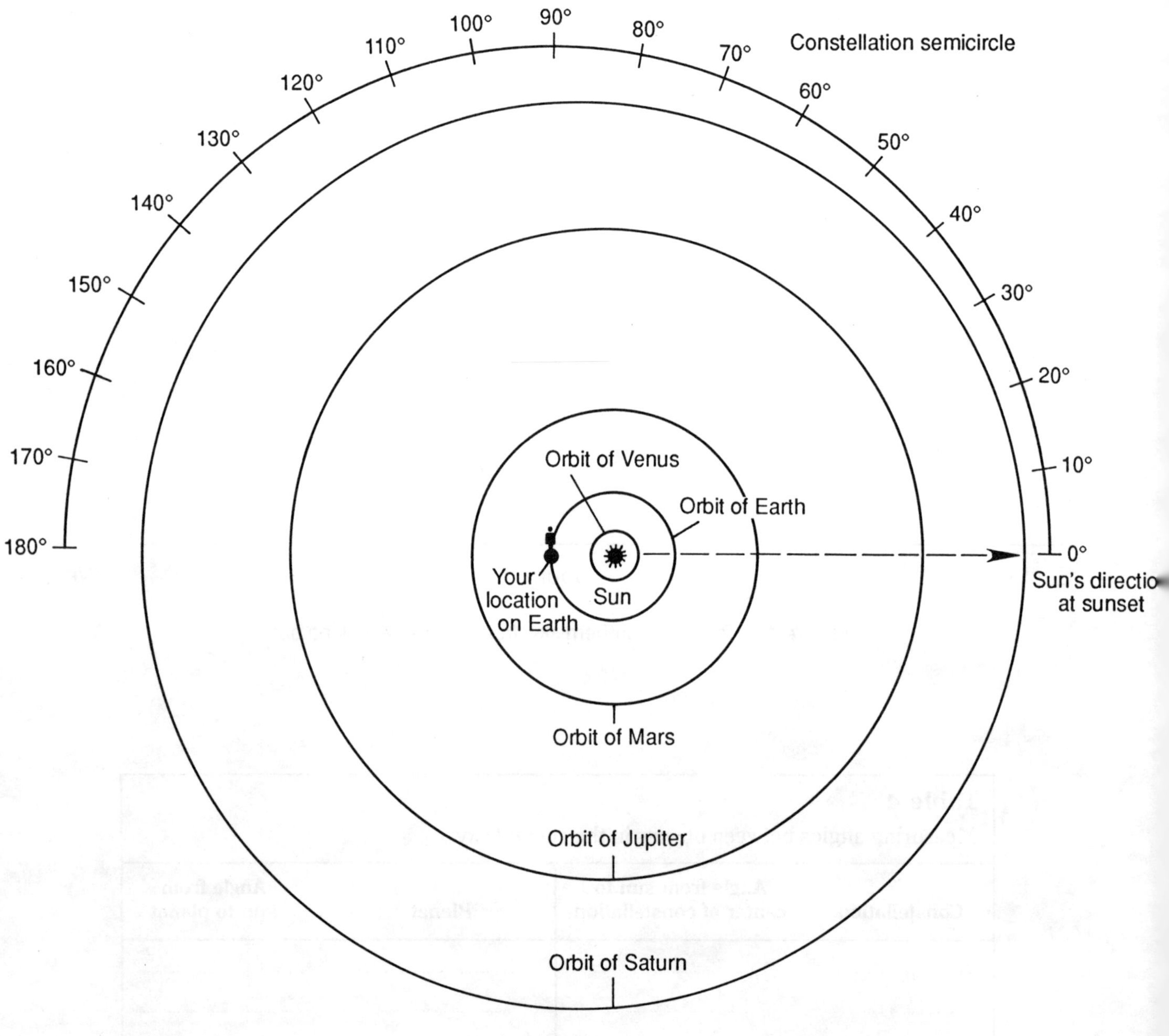

Figure 12. Model map of the solar system, from the outside looking in. (Model is not to scale.)

Name ....
Class ....
Date ....

Observing Our Solar System

Worksheet 5

Table 5
Angles between the sun and sky objects, as measured from two planets.

Sky object	**Angle from sky object to sun, as seen from the planet you landed on** **Planet 1**	**Planet 2** ....
Sun	0°	0°
Venus		
Earth		
Mars		
Jupiter		
Saturn		
Constellation		
Constellation		

Evening sky from planet 1

Planet name ....

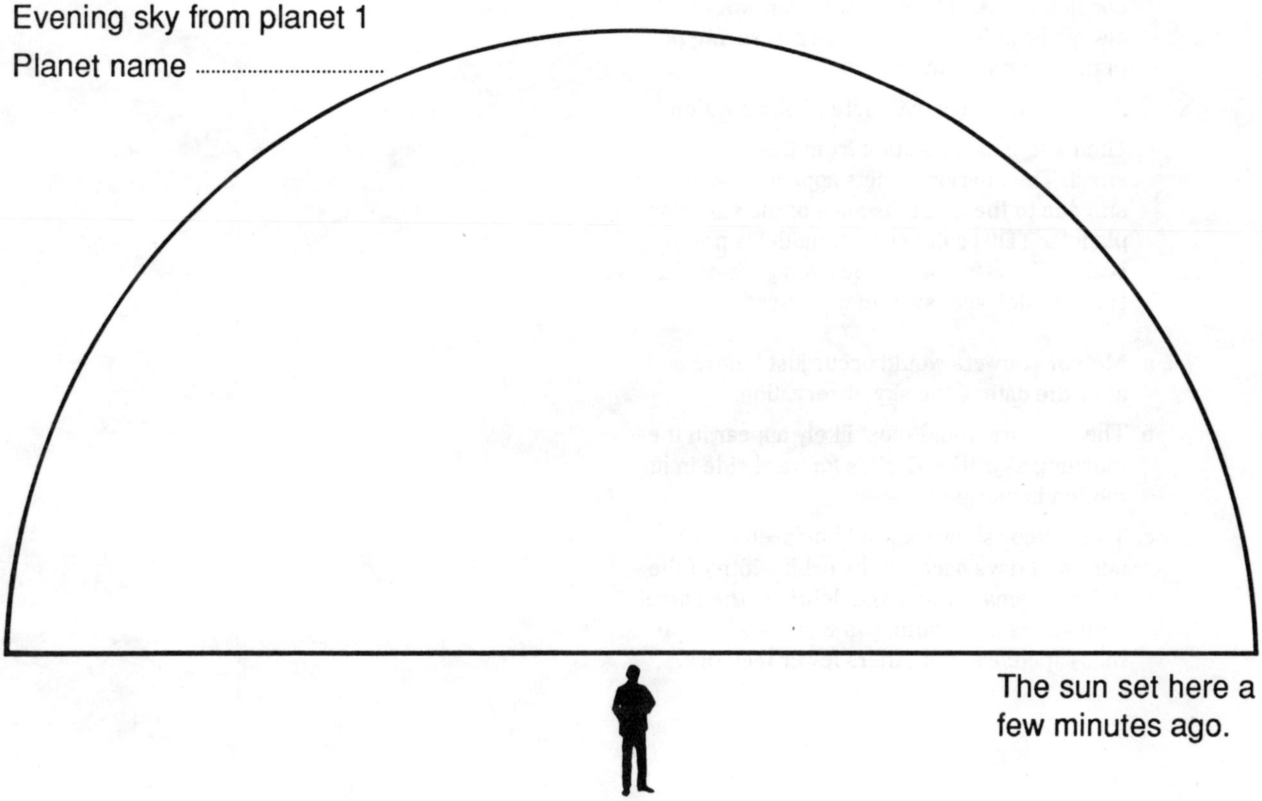

Figure 13. Viewer's location on planet 1.

Answer Key

Questions

1. Students should give the date their observations were made.

2. Angles can be any angular measurement that does not intersect a planet.

3. Answer will vary with date of observation.

4. a. Arrows should point in a counterclockwise direction.

   b-c. Answers will vary with date of observation.

5. a. No, the planets would appear in different constellations. The sun, however, would always be at 0°, the reference measuring point in the solar system.

   b. Answers will vary with date of observation.

   c. Their angle of separation from the sun is small. The inferior planets appear close to the sun due to the great distance of the superior planets. (This solar system model is not to scale. The effect would be even greater if a scale model solar system were used.)

6. a. Meteor showers would occur just before and after the date of the sky observation.

   b. The meteors would most likely appear in the morning sky. The Earth's forward side in its motion is morning.

   c. The meteor showers would be seen over a period of days because the orbit width of the debris is small. Since the debris in the comet path varies at random, some years will have many meteors, and others fewer meteors.

Comparing Size and Density in the Solar System

RANDY R. PATRICK
Southmont Senior High School
Crawfordsville, Indiana

Objectives

To gain an understanding of the distribution of planets and moons in the solar system.

To identify and interpret similarities and differences among the planets and moons by comparing characteristics such as size, density, and composition.

Time required

Two class periods

Materials

No additional materials are needed.

Background

Note: Students will need some preparation for this investigation. See **Additional resources** for a list of required readings for the instructor and students.

Earth is the natural place to begin investigating our solar system. You can learn a lot about it by comparing the planets. Like Earth, Mercury, Venus, and Mars are dense planets, and each has a rocky crust. Thus, the four planets nearest the sun are called **terrestrial** or earthlike.

The planets in the outer solar system are much different. Unlike Earth, Jupiter, Saturn, Uranus, and Neptune have no solid surface. Their thick gaseous atmospheres gradually blend with liquid hydrogen interiors. These planets are similar in that each has a low density and large diameter. They are called the **Jovian** (Jupiterlike) planets. (Jove is another name for Jupiter, the Roman king of the gods.) The most distant planet, Pluto, has a low density, but it is small compared to the Jovian giants.

Scientists hope to discover the process by which the solar system formed by studying its characteristics. From radioactive dating, they have determined that the Earth, moon, and meteorites are about 4.6 billion years old. It appears that our solar system is about the same age.

Our solar system may have developed from a cloud of gas and dust, the **solar nebula.** According to the solar nebula theory, as the cloud contracted, it flattened into a rotating disk. The center of the cloud became the sun, and the planets gradually formed in the disk. Denser planets formed in the inner part of the nebula, where temperatures were so hot that only high-density minerals could form. In the cooler outer regions, gases and dust condensed to form icy material of lower density. (The ice content in the planets seems to increase as the distance from the sun increases). Small solid bodies called planetesimals grew and merged to become protoplanets and eventually planets.

Procedure

Part I. Comparing the planets

Use the data in Table 1 to construct a graph on Worksheet 1, comparing the planets by size and density. Answer the questions after you have completed the graph.

Table 1
Planets of the solar system.

Name	Diameter (km)	Density (g/cm^3)	Surface characteristics	Average distance from the sun (km x 10^6)
Mercury	4,878	5.43	rock	57.9
Venus	12,104	5.24	rock	108.2
Earth	12,756	5.52	rock	149.6
Mars	6,787	3.94	rock	227.9
Jupiter	142,800	1.33	gaseous	778.4
Saturn	120,000	0.70	gaseous	1,423.6
Uranus	51,200	1.30	gaseous	2,866.0
Neptune	48,600	1.76	gaseous	4,488.4
Pluto	2,300	1.1?	ice	5,906.2

Source: *Observer's Handbook - 1990.* (Toronto: Royal Astronomical Society of Canada). Published annually.

Questions

1. Do the planets show any grouping? If so, how many groups are formed?
2. What planets are in each group?
3. Do the planets in these groups have any common characteristics other than size and density? If so, what are these characteristics?
4. Pluto does not seem to fit into any of the groups. What might be the reason for this? How could the distance from the sun cause Pluto to be different from the other planets?
5. Using the graph you have constructed, compare the groups of planets with their distance from the sun. Does this comparison tell you something about the planets and the way they formed? What does this tell you about the temperature in the early solar system?

Part II. Terrestrial planets and major moons

Use the data in Table 2 to construct a graph on Worksheet 2, comparing the terrestrial planets and the major moons of the solar system. Number the moons in Table 2 from 1 to 22; when you plot them on the graph, use numbers instead of dots. Answer the questions after you have completed the graph.

Table 2

Planets and major moons of the solar system.

Name	Diameter (km)	Density (g/cm^3)	Name	Diameter (km)	Density (g/cm^3)
Mercury	4,878	5.43	Dione	1,120	1.4
Venus	12,104	5.24	Tethys	1,060	1.2
Earth	12,756	5.52	Enceladus	500	1.1
Moon	3,476	3.34	Mimas	390	1.2
Mars	6,787	3.94	Uranus		
Phobos	21	~2	Miranda	485	1.26
Deimos	12	~2	Ariel	1,160	1.65
Jupiter			Umbriel	1,110	1.44
Io	3,630	3.55	Titania	1,610	1.59
Ganymede	5,260	1.93	Oberon	1,550	1.50
Callisto	4,800	1.83	Neptune		
Europa	3,140	3.04	Triton	2,720	?
Saturn			Nereid	340	?
Titan	5,150	1.88	Pluto		
Iapetus	1,460	1.2	Charon	1,200	?
Rhea	1,530	1.3			

Source: *Observer's Handbook - 1990.* (Toronto: Royal Astronomical Society of Canada). Published annually.

Questions

1. Do the moons and terrestrial planets appear to group together? If so, list the planets and moons of each group.
2. Give a reason or reasons for this grouping.
3. Describe each of the above groups as to their similarities and differences.
4. Why do the larger moons divide into two distinct groups?
5. Mars and Mercury might form one group by size, but they vary in density by a large amount. What might be the cause of this difference in density?
6. Does the distance from the sun affect the location of a moon or planet on the graph?
7. Does the distance from the sun affect the composition of a moon or planet?
8. Why couldn't you plot the moons of Neptune and Pluto?

Additional resources

Many slide sets that show the most recent views of the planets are available. In addition, your library should have several books showing recent images of the planets and moons.

The following list of books and articles are for the student and the instructor. The selections marked with an asterisk are required reading to complete this investigation.

Books

*Beatty, J. Kelly; O'Leary, Brian; and Chaikin, Andrew, eds. *The New Solar System.* 2d ed. New York: Cambridge University Press, 1981.

Cadogan, Peter H. *The Moon — Our Sister Planet.* New York: Cambridge University Press, 1981.

*Carr, Michael H. *The Surface of Mars.* New Haven: Yale University Press, 1981.

Glass, Billy P. *Introduction to Planetary Geology.* New York: Cambridge University Press, 1982.

Hartmann, William K. *Moons and Planets.* 2d ed. Belmont, Calif.: Wadsworth Publishing Co., 1983.

Morrison, David. *Voyages to Saturn.* NASA SP-451. Washington, D.C.: NASA, 1982.

Morrison, D., and Samz, J. *Voyage to Jupiter.* NASA SP-439. Washington, D.C.: NASA, 1980.

Murray, Bruce G.; Marlin, Michael C.; and Greeley, Ronald. *Earthlike Planets.* San Francisco: W.H. Freeman and Co., 1981.

Journals

Mercury
**Scientific American* (Sept. 1975): p. 58.
Mercury (Sept./Oct. 1984): p. 136.

Venus
Scientific American (July 1981): p. 66.
Astronomy (Sept. 1982): p. 7.
The Planetary Report (Oct./Nov. 1984): p. 12.
**Scientific American* (March 1985): p. 46.

Moon
**Scientific American* (Sept. 1975): p. 92.
The Planetary Report (Feb./March 1982): p. 4.
Astronomy (Oct. 1983): p. 64.
Discover (March 1985): p. 64.

Earth
Scientific American (Sept. 1975): p. 82.
**Scientific American* (June 1981): p. 162.
Science 1982 (April 1982): p. 10.
Astronomy (May 1983): p. 62.
Astronomy (Nov. 1983): p. 62.

Mars
Scientific American (Jan. 1976): p. 32.
**Scientific American* (Feb. 1977): p. 30.
**Scientific American* (March 1978): p. 76.
Science 1981 (July/Aug. 1981): p. 12.
Astronomy (Aug. 1982): p. 60.

Jupiter
Scientific American (May 1976): p. 108.
The Planetary Report (Dec. 1980/Jan. 1981): p. 8.
The Planetary Report (June/July 1981): p. 8.

	Scientific American (Dec. 1981): p. 90.
	**Science 1983* (March 1983): p. 6.
	Mercury (Jan./Feb. 1984): p. 23.
	Mercury (Nov./Dec. 1984): p. 169.
Saturn	*The Planetary Report* (Feb./March 1981): p. 5.
	The Planetary Report (Oct./Nov. 1981): p. 3.
	Science 1981 (Nov. 1981): p. 41.
	**Scientific American* (Jan. 1982): p. 100.
	The Planetary Report (Oct./Nov. 1983): p. 3.
	Astronomy (Nov. 1984): p. 60.
Uranus	*Mercury* (March/April 1983): p. 58.
	Astronomy (Sept. 1985): p. 6.
	Sky and Telescope (Jan. 1986): p. 11.
	**Sky and Telescope* (April 1986): p. 333, 338.
Neptune	*Astronomy* (Sept. 1982): p. 62.
	Science 1983 (Oct. 1983): p. 7.
	Science 1984 (Oct. 1984): p. 12.
Pluto	**Science 1982* (Sept. 1982): p. 62.
	Astronomy (April, 1983): p. 62.
	The Planetary Report (Dec. 1984/Jan. 1985): p. 11.
	**Scientific American* (June 1990): p. 50.

*Required reading for students and instructor.

Comparing Size and Density in the Solar System

Worksheet 1

Name..........

Class..........

Date..........

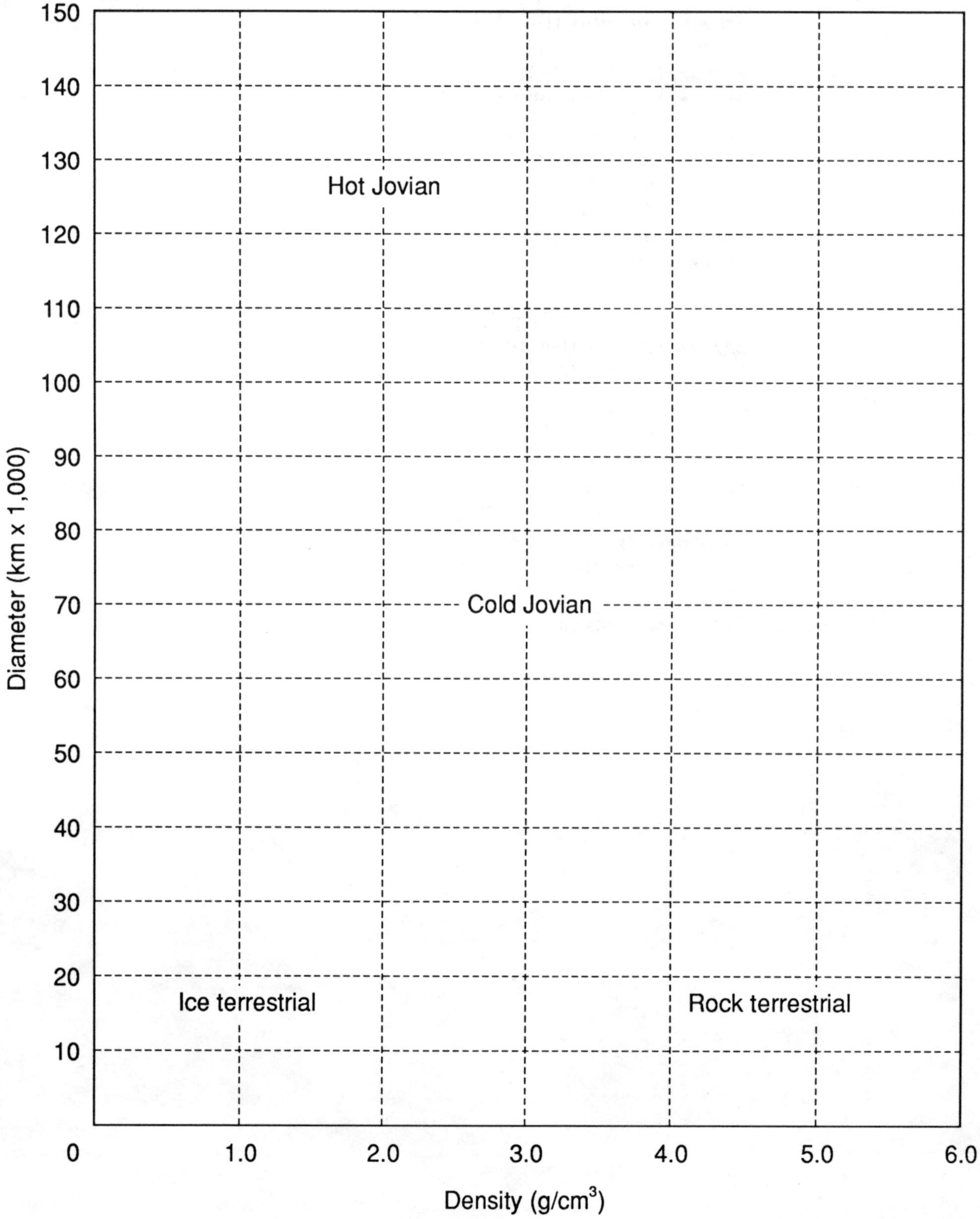

Figure 1. Graph of diameter vs. density of planets in our solar system.

Name ....

Class ....

Date ....

Comparing Size and Density in the Solar System

Worksheet 2

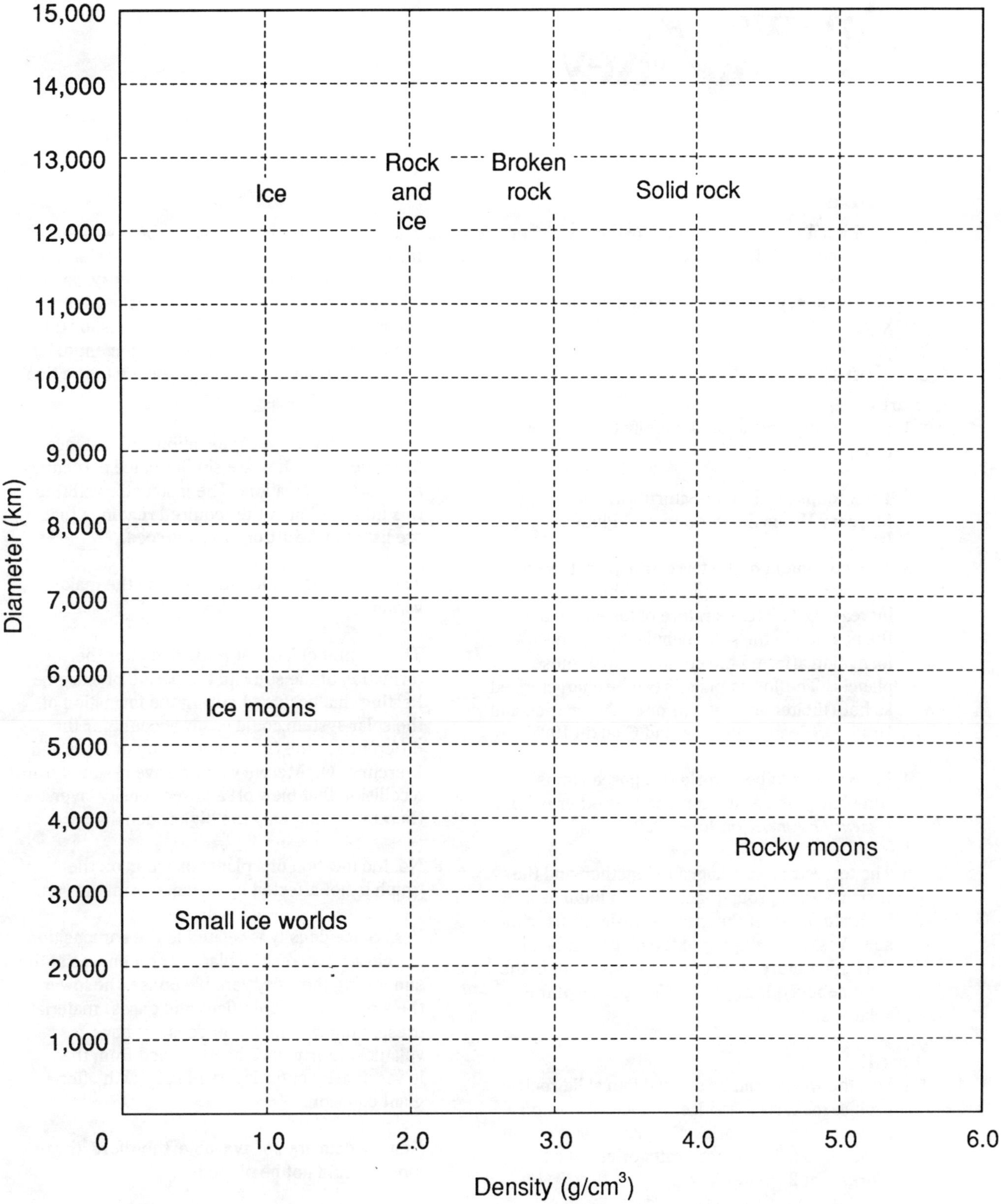

Figure 2. Graph of diameter vs. density of terrestrial planets and major moons.

Answer ✓ Key

Procedure

Additional readings assigned by the instructor would be helpful for the class.

See the next pages for the worksheet answer keys.

Questions

Part I.

1. Yes, the students should identify three or four groups.

2. The groups are Jupiter-Saturn, Uranus-Neptune, Mercury-Venus-Earth-Mars, and Pluto.

3. The ice content of the terrestrial planets seems to increase as the distance from the sun increases. The temperature of formation and the position in the solar nebula during the formation affected the composition of these planets. The Jovian planets can be characterized as hot (Jupiter and Saturn) or cold (Neptune and Uranus) (Beatty, O'Leary, and Chaikin 1981).

4. Pluto seems to be a world composed almost completely of ice, and it resembles other moons (*Scientific American*, June 1990, p. 50).

5. The terrestrial worlds group together, and the Jovian worlds group together with Pluto as a separate planet at the greatest distance from the sun. Yes, the grouping of the planets indicates the temperature and composition present in the solar nebula during the earliest period of the solar system.

Part II.

1. Yes, the worlds can be divided into solid rock, broken rock, rock and ice, and ice. Although their composition varies, the planets and moons all seem to fall into these categories. See the Worksheet 2 answer key.

2. The temperature in the solar nebula was the major force leading to the final composition of the moons and planets (Beatty, O'Leary, and Chaikin 1981). The moons also make a large group and a small group. The reasons for this grouping seem to relate to the placement of planetismals and their coalescence into the final planets and moons.

3. The groups vary in composition and distance from the sun. They are similar in ice percentage and heat of formation. The instructor will find this information in the required readings that are listed in **Additional resources**.

4. The percentage of ice seems to be the major variance.

5. The amount of ice that remained after the formation of the sun and the period of extreme heating that occurred during the formation of the solar system could easily account for the difference in the final composition of Mars and Mercury. Or, Mercury could have resulted from a collision that blew off a lower density layer, thereby leaving only the higher density core.

6. No, the location of a planet or moon on the graph is not affected by distance from the sun.

7. Yes. Since density is related to the composition, the closer a terrestrial planet or moon was to the sun during the early variable phase, the lower the amount of volatile (ices and gases) materials present today. The moon does not have volatiles. It may have been formed from the lower density outer layers of the Earth after a giant collision.

8. Density data are not available; therefore, these moons could not be plotted.

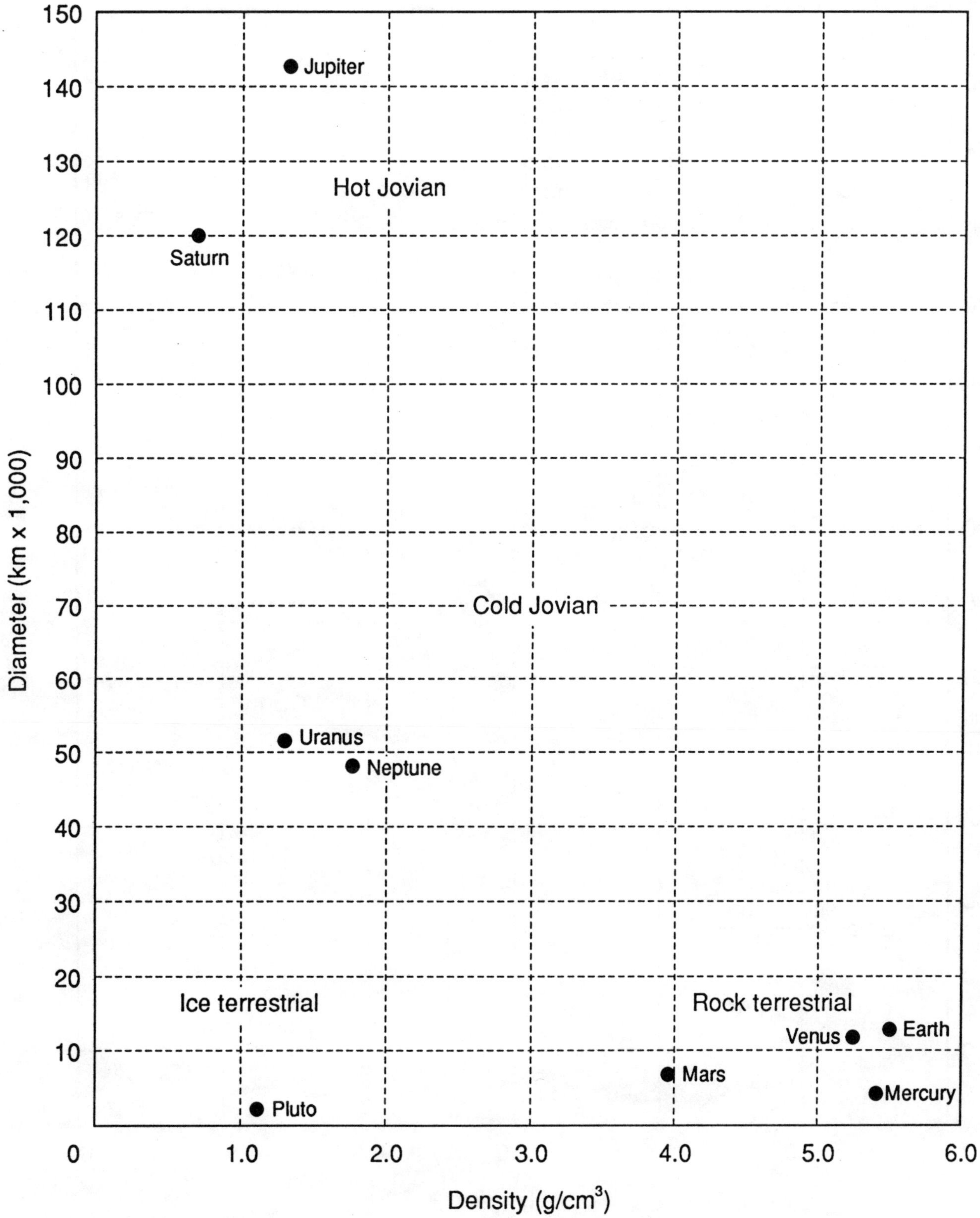

Figure 1. Graph of diameter vs. density of planets in our solar system.

Answer Key

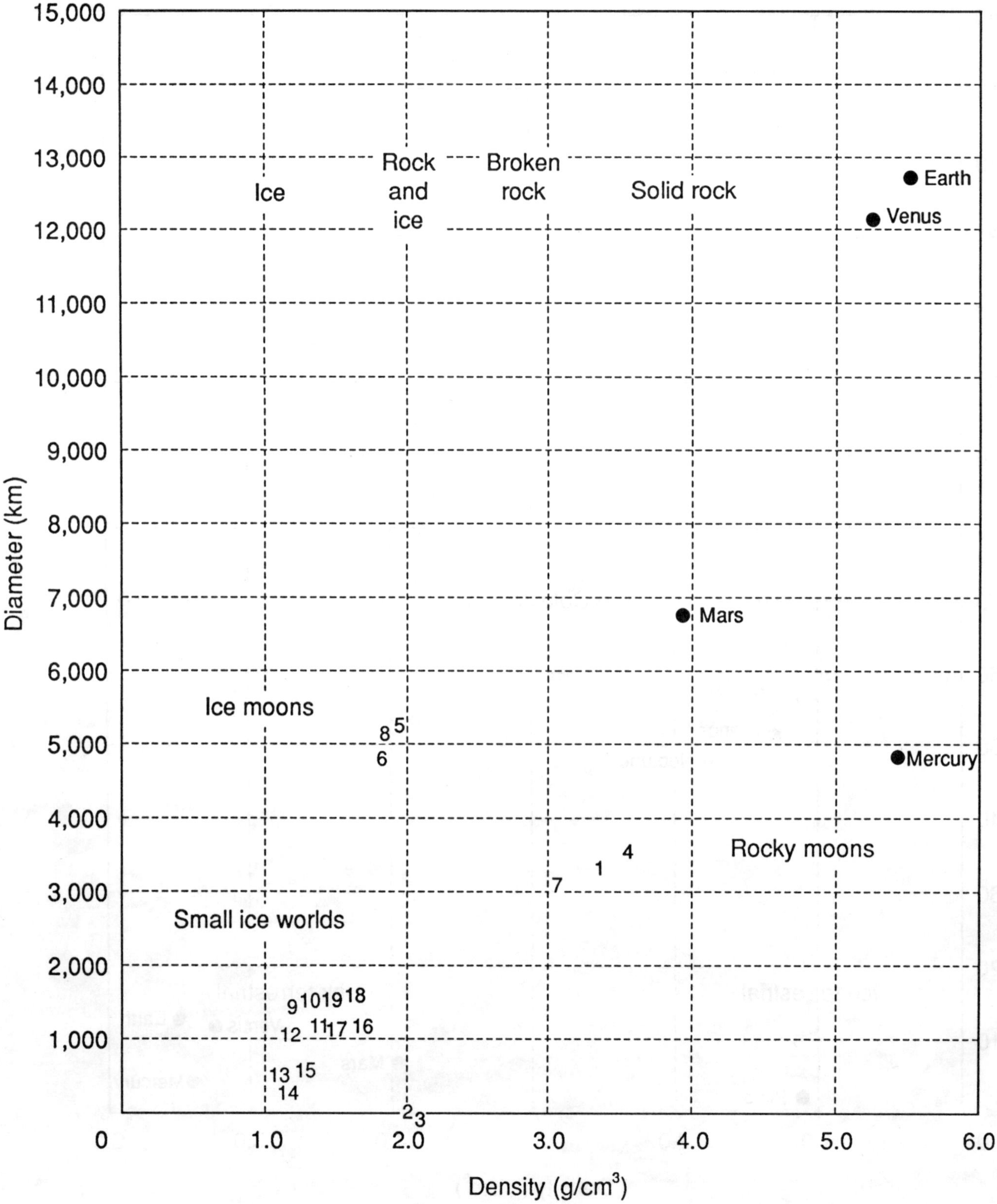

Figure 2. Graph of diameter vs. density of terrestrial planets and major moons.

Phases of the Moon

Paul Johnston
Emporia State University
Emporia, Kansas

Objectives

To understand the relationship between phases of the moon and the position of the moon relative to the sun and the Earth.

To identify phases of the moon at different positions in its orbit.

To determine the time of day the moon rises in its various phases.

To determine the phase the Earth would appear to be in when viewed from the moon at different positions in the moon's orbit.

Time required

One class period

Materials

Note: For the items listed below, see **Instructions for Assembling Astronomy Materials**, p. 211.

per class or group
Pegboard base
2 polystyrene spheres on golf tees to represent the Earth and the moon
Overlay 1 (p. 214-215)

Background

As the moon revolves around the Earth, its position continually changes relative to the Earth and the sun. The amount of the sunlit portion of the moon that we can see from Earth varies from day to day. Normally, this illuminated part of the moon is all that we are able to see. The unlighted part of the moon is dark and invisible, because it blends into the night sky.

Phases of the moon are the recurring changes in its illumination and apparent shape. You have probably seen the full moon or the crescent moon many times. A gibbous moon describes a phase in which more than half, but not all, of the disk is illuminated.

Procedure

Part I. Phases of the moon

1. Put Overlay 1, showing eight positions in the moon's orbit, on the pegboard.

2. Put the Earth in the center of the moon's orbit and the moon in position 1.

3. The white halves of the Earth and moon represent their daylight sides. Aim the daylight sides of the Earth and moon toward the edge of the board that shows rays of light from the sun.

4. Move the moon in its orbit in a counterclockwise direction around the sun from position 1 through position 8. Be sure to keep the daylight side of the moon aimed in the direction of the sun's rays at all times. At each numbered position, sight across the Earth toward the moon, and observe the shape of the white portion of the moon that you see. The shapes illustrate the phase of the moon at various positions in its orbit when it is viewed from the Earth.

5. Match the eight phases of the moon, shown on Worksheet 1, Figure 1, with the numbered positions on the model of the moon's orbit. Write the appropriate number of the moon's position under the name of a phase in Figure 1.

Part II. Times of moon rise

1. The positions labeled *A-H* around the equator on the Earth represent different times of the day. Assume equal hours of night and day. The letter *C* is in the center of the day half of the Earth. The letter *E* is on the line that separates day from night. Determine the time of day that is indicated by each of the lettered positions. Keep in mind that as seen from above, the Earth rotates counterclockwise on its axis. Record the times on Worksheet 1, Table 1.

2. The time of day that the moon rises depends on which phase it is in. Aim the daylight side of the Earth toward the sun. Put the moon in the new crescent position with its daylight side also aimed toward the sun.

3. Hold the board so that the Earth blocks your view of the moon. Slowly peek around the right side of the Earth until you just begin to see the edge of the moon. Notice that when the moon first begins to come into view, your line of sight grazes the letter *B* position on the Earth. You would have to be standing at position B on the Earth to see the moon rise when it is in the new crescent phase. What time of day is it? Record the time on Worksheet 2, Table 2.

4. Determine the time of moon rise for each of its phases, and record the times in Table 2.

Part III. Phases of the Earth as seen from the moon

When astronauts are on the moon, they see the Earth going through phases much as we see the phases of the moon from the Earth. Sight across the moon from each of its eight positions, and determine the phase the Earth would be in as seen from the moon. Record the phases on Worksheet 2, Table 3.

Additional resources

Alexander, Peter, et al. *Earth Science.* Morristown, N.J.: Silver Burdett and Ginn, 1987.

Flohr, Terry, ed. *Focus on Earth Science.* Columbus: Merrill Publishing Co., 1984.

Ramsey, William L.; Phillips, Clifford R.; and Waterpough, Frank M. *Modern Earth Science.* New York: Holt, Rinehart and Winston, 1983.

Worksheet 1

Name ....
Class ....
Date ....

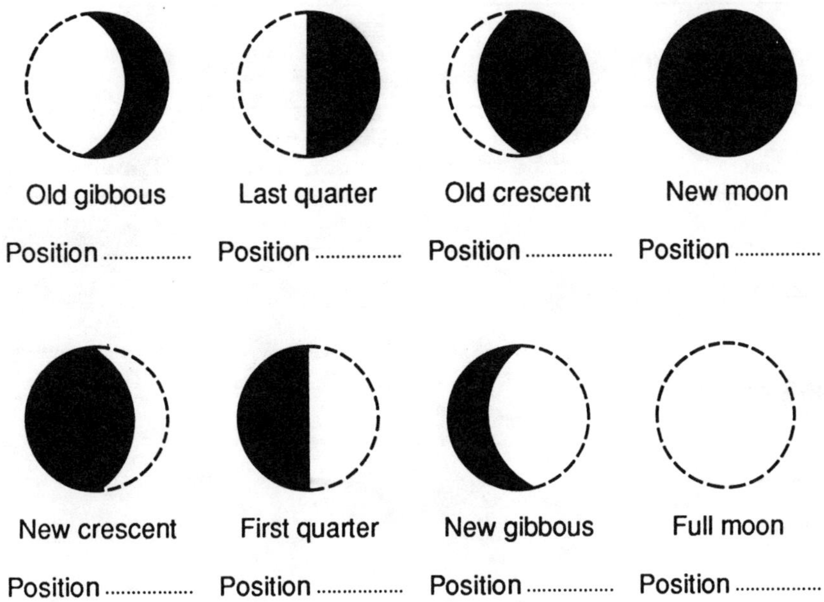

Figure 1. Phases of the moon.

Table 1
Time of day at different Earth positions.

Position on Earth sphere	Time of day
A	....
B	....
C	....
D	....
E	....
F	....
G	....
H	....

Worksheet 2

Name ....

Class ....

Date ....

Table 2
Times of moon rise.

Moon phase	Position on the Earth where one must be to see the moon rise	Time of moon rise
New crescent	....	....
First quarter	....	....
New gibbous	....	....
Full moon	....	....
Old gibbous	....	....
Last quarter	....	....
Old crescent	....	....
New moon	....	....

Table 3
Phases of the moon and the Earth.

Moon position	Moon phase as seen from the Earth	Earth phase as seen from the moon
1	....	....
2	....	....
3	....	....
4	....	....
5	....	....
6	....	....
7	....	....
8	....	....

Answer Key

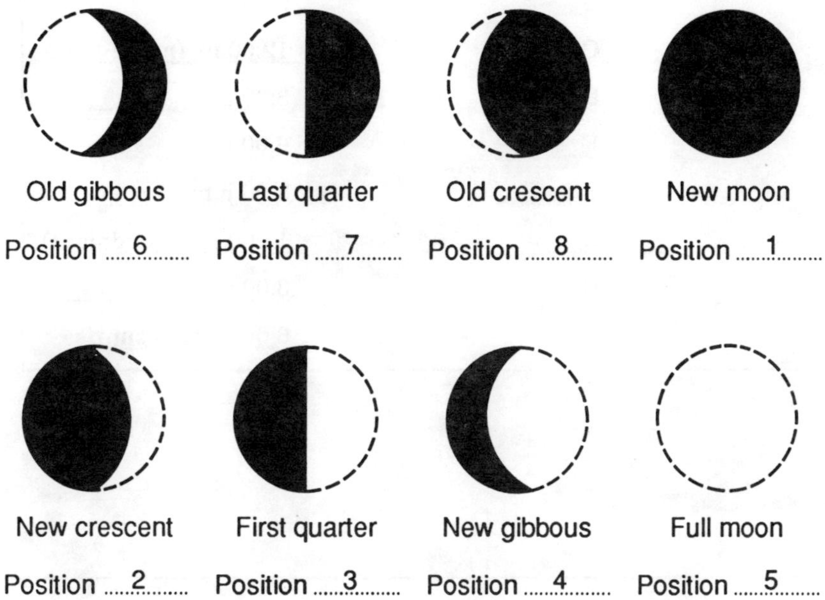

Figure 1. Phases of the moon.

Table 1
Time of day at different Earth positions.

Position on Earth sphere	Time of day
A	6:00 a.m. — sunrise
B	9:00 a.m.
C	12:00 m. (noon)
D	3:00 p.m.
E	6:00 p.m. — sunset
F	9:00 p.m.
G	12:00 p.m. (midnight)
H	3:00 a.m.

Note: The Latin word ***meridies*** means noon, and its abbreviation is "m." ***Ante meridiem*** (a.m.) refers to time before noon, and ***post meridiem*** (p.m.), to time after noon. Midnight (12:00 p.m.) is time after noon.

Answer Key

Table 2
Times of moon rise.

Moon phase	Position on the Earth where one must be to see the moon rise	Time of moon rise
New crescent	B	9:00 a.m.
First quarter	C	12:00 m. (noon)
New gibbous	D	3:00 p.m.
Full moon	E	6:00 p.m. — sunset
Old gibbous	F	9:00 p.m.
Last quarter	G	12:00 p.m. (midnight)
Old crescent	H	3:00 a.m.
New moon	A	6:00 a.m. — sunrise

Table 3
Phases of the moon and the Earth.

Moon position	Moon phase as seen from the Earth	Earth phase as seen from the moon
1	New moon	Full Earth
2	New crescent	Old gibbous
3	First quarter	Last quarter
4	New gibbous	Old crescent
5	Full moon	New Earth
6	Old gibbous	New crescent
7	Last quarter	First quarter
8	Old crescent	New gibbous

Solar and Lunar Eclipses

PAUL JOHNSTON
Emporia State University
Emporia, Kansas

Objective

To determine how many solar eclipses and lunar eclipses occur in one year.

Time required

One class period

Materials

Note: For the items listed below, see **Instructions for Assembling Astronomy Materials**, p. 211.

per class or group
Pegboard base
3 polystyrene spheres on golf tees to represent the Earth, moon, and sun
2 dowels: 1/4-in. diameter, 3 in. long
Block of wood: 2 in. x 2 in. x 3 in.
Overlay 2 (p. 216-217)

Background

The moon takes about four weeks to orbit the Earth. During the orbit, the part of the moon that faces the sun is illuminated. The apparent shape of this sunlit portion varies depending on the moon's position in relation to the sun. These varied and recurring shapes are called phases and are shown in Figure 1.

We see a new moon when it is between the Earth and the sun because the dark half of the moon is aimed toward the Earth. A **solar eclipse** occurs when the moon blocks our view of the sun while moving between it and the Earth.

When the moon is on the side of the Earth opposite to that of the sun, the sun illuminates the moon, and it appears full. A **lunar eclipse** occurs when the moon moves into the Earth's shadow during the full phase.

If the Earth, moon, and sun stayed on the same plane of revolution, we would see a solar eclipse every new moon and a lunar eclipse every full moon. The result would be about 12 solar and 12 lunar eclipses each year. However, there is about five degrees difference between the moon's plane of revolution and that of the Earth. As a result, most new and full moons occur when the moon is either above the Earth's plane of revolution or below it.

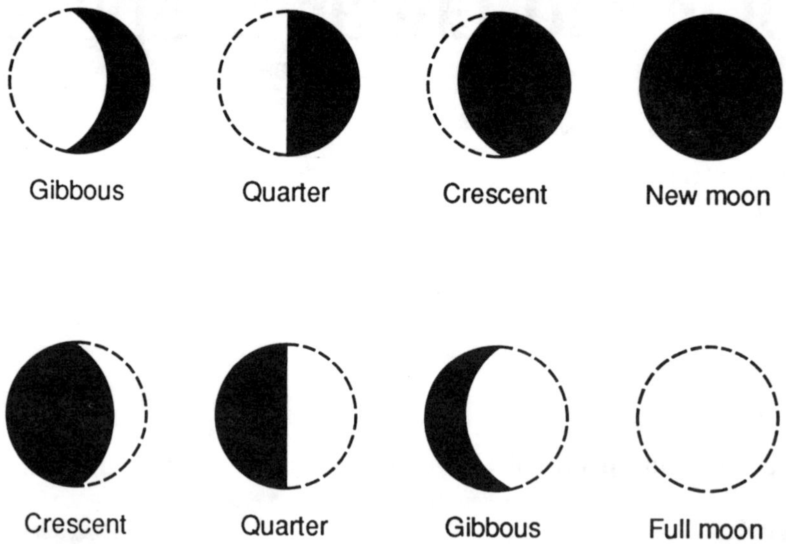

Figure 1. Phases of the moon. Notice the difference between the crescent phases and the gibbous phases: in the crescent phases, more than half, but not all, of the disk is dark; in the gibbous phases, more than half, but not all, of the disk is illuminated.

Since the moon orbits on a plane that is inclined to that of the Earth, it is above the Earth's plane for about two weeks. Then the moon crosses below Earth's plane and remains there for two weeks before crossing above again. We refer to the place that the moon crosses Earth's plane of revolution as a **node**. The moon is on a node twice a month. It is on a descending node as it moves below Earth's plane of revolution. Two weeks later, an ascending node occurs when the moon crosses above our plane.

The relationship between nodes and phases of the moon determines when a solar or lunar eclipse will occur. To cause a solar eclipse, the moon must be in new phase, and at the same time, it must be on a node. For a lunar eclipse to occur, the moon must be full as it reaches one of the nodes.

Procedure

1. Put Overlay 2, showing eight positions in the moon's orbit, on the pegboard.

2. Insert the Earth in the center of the moon's orbit and the moon in position 3. Place the sun in a $3/16$-in. hole in one end of a 2 in. x 2 in. x 3 in. block of wood.

3. Insert a dowel in a $1/4$-in. hole on each leg of the pegboard so that the plane of revolution of the moon is slanted to that of the Earth, as shown in Figure 2. Put the sun beside the board so that the moon is between it and the Earth. The plane of revolution of the Earth is parallel to the table top at the level of the sun and the Earth. The plane of revolution of the moon is parallel to the pegboard, the Earth-moon board, and is inclined about 5° to the plane of revolution of the Earth.

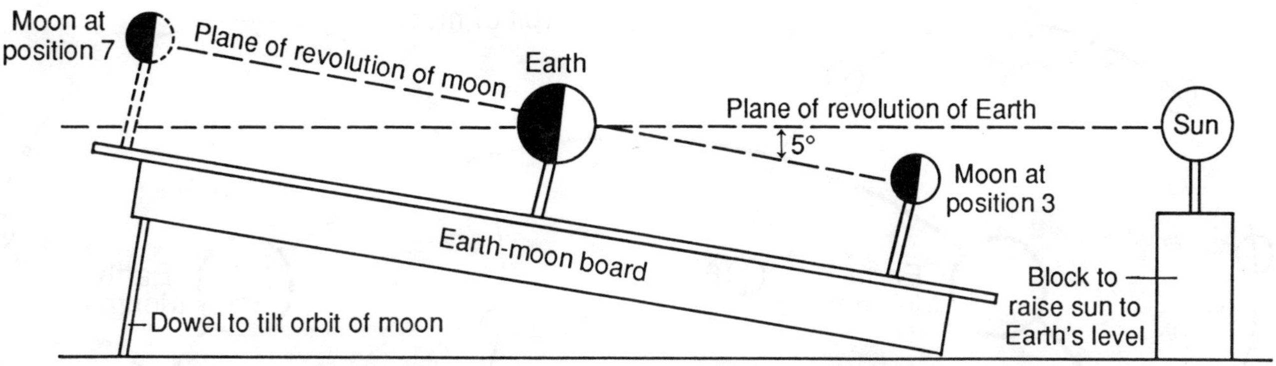

Figure 2. Diagram showing placement of the Earth, moon, and sun and the inclination of the Earth-moon board.

4. Figure 3 shows the position of the Earth, moon, and sun for four different months of the year. Each of the eight moon positions shown for each month represents a different phase. By moving the Earth-moon board around the sun, you can simulate one year's motion.
   a. Lay Figure 3 on the table, and put the sun in the center of the figure. Following the diagram for month I in Figure 3, position the Earth-moon board next to the sun with moon position 3 closest to the sun. Figure 2 shows the setup.
   b. Determine the phase of the moon at each of the eight numbered positions in its orbit for month I. Write the phases under "Month I" in Table 1 on the worksheet.

5. Follow the diagram for month II in Figure 3 to position the Earth-moon board next to the sun with moon position 5 closest to the sun. Determine the phase of the moon at each of the eight positions in its orbit for month II. Write the phases under "Month II" in Table 1. Notice that during month I, the diagram shows that moon position 3 is between the sun and the Earth, but in month II, moon position 5 is between the sun and the Earth. Although moon position 3 is new moon at month I, it is a different phase in month II.

6. Follow the Figure 3 diagrams for months III and IV to position the Earth-moon board in relation to the sun. Determine the phase of the moon for each of the eight positions for both months. Write the phases in Table 1.

   Because the Earth-moon board is tilted, some of the moon's numbered positions are above the level of the Earth's rotational plane and others are below it. At two positions, the moon and the Earth are on the same level. These are the nodes.

Questions

1. Where on the moon's orbit do the ascending and descending nodes occur? Write the numbered position for each. Notice that the nodes occur at the same positions each month.
2. For a solar eclipse to occur, the moon must be between the Earth and the sun (new moon), and the moon must be on one of the nodes. Look at Table 1, showing which phase the moon was in for each numbered position during the four months. Locate the places where the new moon occurred on a node. How many times during the year was the new moon on a node?

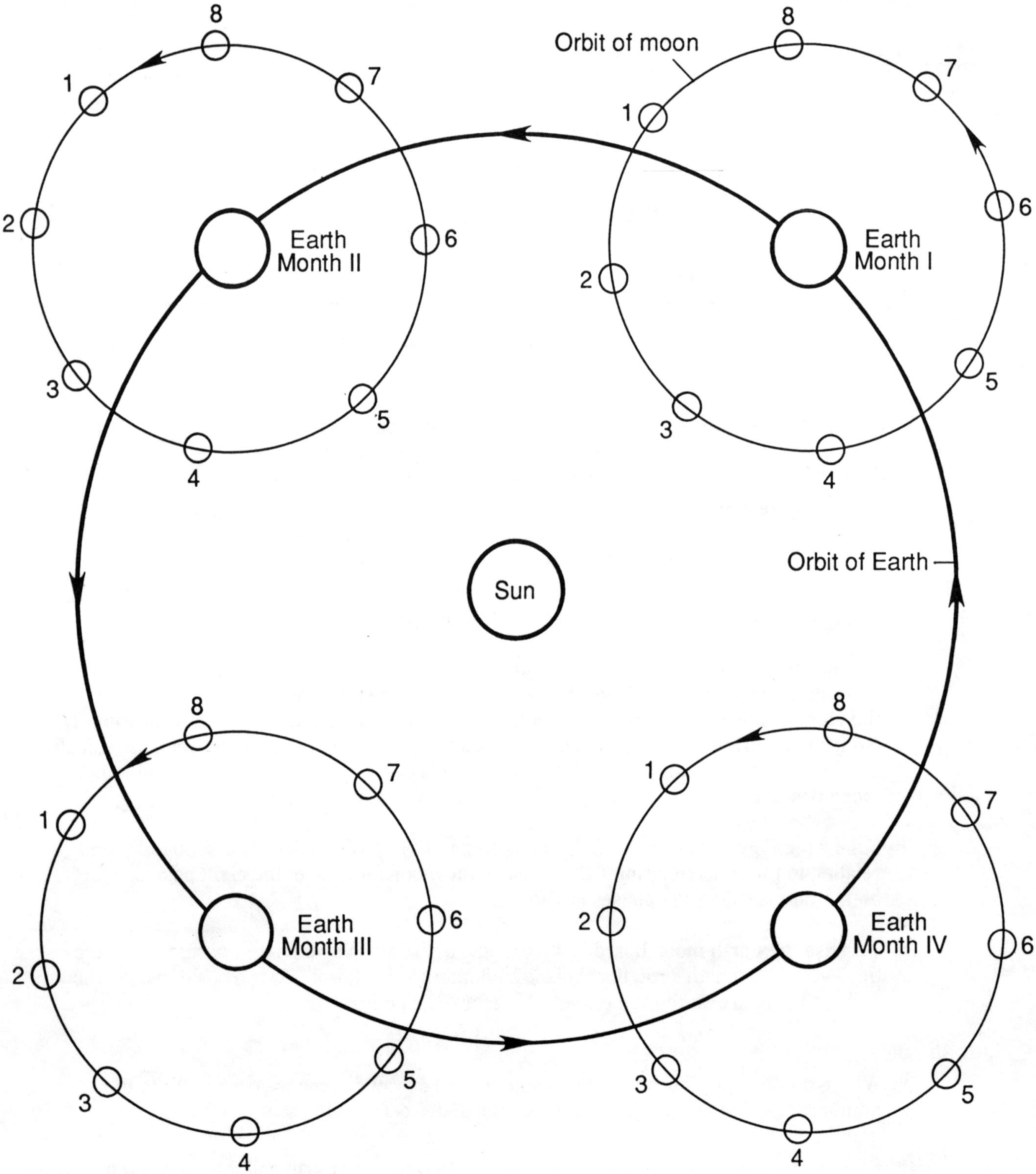

Figure 3. Position of the Earth, moon, and sun for four different months of the year.

3. During which of the months I-IV did a new moon occur on a node?
4. For a lunar eclipse to occur, the moon must be on the side of the Earth opposite to that of the sun (full moon), and the moon must be on one of the nodes. Look at your table again, and locate the places where a full moon occurs on a node. How many times during the year did a full moon occur on a node?
5. During which of the months I-IV did a full moon occur on a node?
6. Would you expect to see a lunar eclipse and a solar eclipse during the same month?
7. When we on Earth see a lunar eclipse, what would astronauts on the moon see?
8. What would astronauts on the moon see when we see a solar eclipse on Earth?
9. If the plane of revolution of the moon were parallel to the plane of revolution of the Earth, would we see eclipses more often or less often?

Additional resources

Alexander, Peter, et al. *Earth Science.* Morristown, N.J.: Silver Burdett and Ginn, 1987.

Flohr, Terry, ed. *Focus on Earth Science.* Columbus: Merrill Publishing Co., 1984.

Ramsey, William L.; Phillips, Clifford R.; and Waterpough, Frank M. *Modern Earth Science.* New York: Holt, Rinehart and Winston, 1983.

Worksheet

Name ....

Class ....

Date ....

Table 1
Phases of the moon at different positions in its orbit.

Moon position	Month I	Month II	Month III	Month IV
1	....	....	....	....
2	....	....	....	....
3	....	....	....	....
4	....	....	....	....
5	....	....	....	....
6	....	....	....	....
7	....	....	....	....
8	....	....	....	....

Answer Key

Procedure

The worksheet answer key is shown below.

Questions

1. An ascending node occurs at moon position 5. A descending node occurs at moon position 1.

2. A new moon occurred on a node twice.

3. A new moon occurred on a node during months II and IV.

4. A full moon occurred on a node twice.

5. A full moon occurred on a node during months II and IV.

6. Yes, if the full moon occurs on a node, two weeks later the new moon will also be on a node. Both new and full phases and ascending and descending nodes occur during the same two-week interval. Lunar and solar eclipses should be visible somewhere on the Earth within about two weeks of one another.

7. Astronauts on the moon would see a solar eclipse when we see a lunar eclipse on Earth. For us to see a lunar eclipse, the Earth must be between the moon and the sun so that Earth's shadow falls on the moon. This alignment would cause the Earth to block out the sun for viewers on the moon.

8. Astronauts on the moon would see an Earth eclipse (moon's shadow on the Earth) while we see a solar eclipse. For us to see a solar eclipse, the moon must be between the Earth and the sun. This alignment would cause the moon's shadow to fall on the Earth. Such an event would not be very spectacular, because the moon's shadow would only appear as a small dot on the Earth.

9. Eclipses would occur more often if the plane of revolution of the moon were parallel to Earth's. We would have a solar eclipse every new moon and a lunar eclipse every full moon.

Table 1
Phases of the moon at different positions in its orbit.

Moon position	Month I	Month II	Month III	Month IV
1	Last quarter	Full moon	First quarter	New moon
2	Old crescent	Old gibbous	New gibbous	New crescent
3	New moon	Last quarter	Full moon	First quarter
4	New crescent	Old crescent	Old gibbous	New gibbous
5	First quarter	New moon	Last quarter	Full moon
6	New gibbous	New crescent	Old crescent	Old gibbous
7	Full moon	First quarter	New moon	Last quarter
8	Old gibbous	New gibbous	New crescent	Old crescent

Geocentric Theory vs. Heliocentric Theory for Venus

PAUL JOHNSTON
Emporia State University
Emporia, Kansas

Objective

To demonstrate that the planet Venus revolves around the sun and that the planet, therefore, cannot be following the motions required of it by the geocentric theory.

Time required

One class period

Materials

Note: For the items listed below, see **Instructions for Assembling Astronomy Materials**, p. 211.

per class or group
Pegboard base
3 polystyrene spheres on golf tees to represent the Earth, Venus, and sun
Overlays 3 and 4 (p. 218-221)

Background

Before the development of the telescope, people did not realize that the planets were visible because they reflected light from the sun and that planets go through phases much like those of the moon. Without this knowledge, there was very little to dispute the geocentric (Earth-centered) model of the solar system.

According to the **geocentric theory**, the sun, moon, and planets revolved clockwise around the Earth. Each

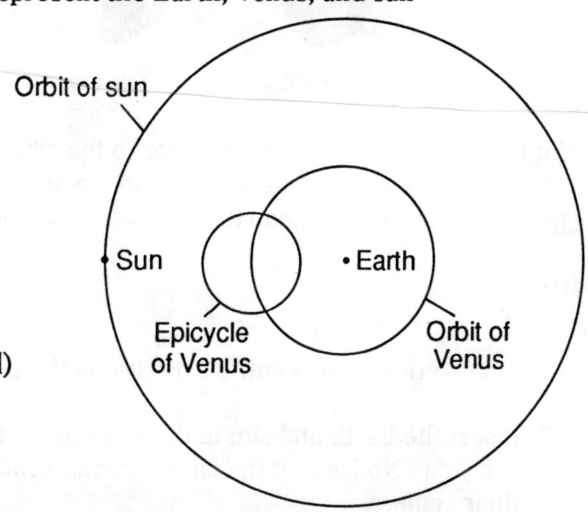

Figure 1. Illustration of the geocentric theory.

planet supposedly also moved in a smaller circle called an epicycle. The center of the epicycle moved along a larger circle or orbit around the Earth (Figure 1).

Any theory regarding the motion of the planets had to explain the fact that Venus was visible only shortly before sunrise in the east or a little after sunset in the west. It was never seen in the part of the sky opposite to the side where the sun was, for example, overhead at midnight. The explanation was that the center of the epicycle Venus followed was on a line between Earth and the sun. This position kept Venus from moving very far to either side of the sun as seen from the Earth.

As astronomers made more accurate observations and collected more data, they found the geocentric theory increasingly difficult to use to predict planetary motion. In 1543, Copernicus, a Polish astronomer, proposed the **heliocentric theory**, the sun-centered model of the solar system, which is now accepted.

To compare the geocentric and heliocentric models, you will need to be able to recognize phases such as the moon goes through as it orbits around the sun. Like the moon, other celestial bodies that are illuminated by the sun go through similar phases. Figure 2 shows eight phases of the moon.

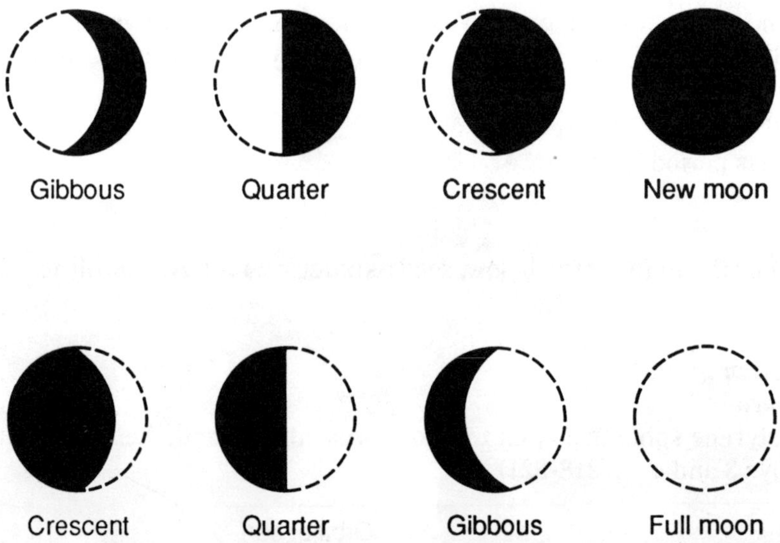

Figure 2. Phases of the moon. Notice the difference between the crescent phases and the gibbous phases: in the crescent phases, more than half, but not all, of the disk is dark; in the gibbous phases, more than half, but not all, of the disk is illuminated.

Procedure

Part I. Geocentric theory

1. Put Overlay 3 (geocentric overlay) on the pegboard.

2. Insert the Earth and sun in their designated locations. Insert Venus in position 1 on its epicycle. Notice that the large circular orbits of the sun and Venus have the Earth at their centers.

3. The white halves of the Earth and Venus represent their daylight sides. Move Venus around its epicycle from position 1 through position 8, keeping the white (lighted) side aimed toward the sun.

4. Sketch on the worksheet, Figure 4, the phase that Venus appears to be in at each of the eight positions. The lighted part is all that you would see; the dark half would blend in with the night sky and not be observed. Notice that in positions 4 and 8, Venus would be difficult to see. The dark half is aimed toward the Earth, and the viewer would be looking into the sun. Given its distance from the Earth, Venus would appear as a tiny black dot on the face of the sun.

5. Write the name of each phase on the dotted lines under your sketches. Refer to Figure 2 for phase names.

Part II. Heliocentric theory

1. Replace Overlay 3 with Overlay 4 (heliocentric overlay) on the pegboard.

2. Insert the Earth and sun in their designated locations. Insert Venus in position 1 on its orbit. Notice that the sun is in the center of the orbits of Venus and the Earth. The two planets revolve around the sun in a counterclockwise direction.

3. Move Venus in its orbit from position 1 through position 8, keeping the white (lighted) side aimed toward the sun. Sketch on the worksheet, Figure 5, the phase that Venus appears to be in at each of the eight positions.

4. Write the name of each phase on the dotted lines under your sketches.

Questions

1. Compare your sketches of Venus using the geocentric model and the heliocentric model with the sketches in Figure 3. Figure 3 sketches show how Venus actually appeared in the night sky. Which theory regarding the motions of the planets do you think is correct? Explain.

Figure 3. Series of sketches that show how Venus appeared when viewed with a telescope on different evenings over a period of several months. Notice that Venus not only appears to change shape (phase), but it also appears to change size.

2. The positions labeled *A-H* around the equator on the Earth represent different times of the day. The letter *C* is in the center of the day half, so it is noon. The letter *G* is in the center of the night half, so it is midnight. The letters *E* and *A* are on the day-night boundary. Which letter represents sunrise, and which represents sunset? Remember that the Earth rotates counterclockwise on its axis. Assume equal hours of night and day.
3. Hold the board so that the black half of the Earth faces you, to simulate your standing in the night part of the Earth. By peeking around the right side of the Earth — your line of sight grazes the letter *A* on the Earth — you can see Venus when it is in positions 5, 6, or 7 in its orbit. If you peek around the left side of the Earth — your line of sight grazes the letter *E* — you can see Venus when it is in positions 1, 2, or 3. At which positions would Venus be visible from Earth in the early morning hours? At which positions would it appear in the evening sky?
4. Study the sketches representing actual observations of Venus (Figure 3). Notice that Venus not only changes shape, but it also changes size. On the heliocentric overlay (Overlay 4), where does Venus appear to be the smallest in its orbit? Write the three numbered positions. Why does Venus look smaller to us when it is in this part of its orbit than when it is at other positions?
5. At which of the numbered positions on the board would Venus appear to be the brightest?
6. Sight across Venus toward the Earth from each of Venus's numbered positions, and observe the phase that the Earth would be in as seen from Venus. Does the Earth ever appear to be in a crescent or quarter phase? Why?
7. Mars is farther from the sun than the Earth, so Mars' orbit is outside the Earth's orbit. Would you expect Mars to exhibit crescent or quarter phases as seen from the Earth?
8. Consider whether planets orbit inside or outside a viewer's orbit. Write a general statement about which of the planets would show all of the phases (crescent, quarter, gibbous, full, and new) and which would always appear to be full or nearly full.

Additional resources

Alexander, Peter, et al. *Earth Science.* Morristown, N.J.: Silver Burdett and Ginn, 1987.

Atkin, J. Myron, and Wyatt, Stanley P., Jr. *The Universe in Motion.* Book 2. The University of Illinois Astronomy Program. New York: Harper and Row Publishers, 1969.

Redfield, David D., and Snyder, William R. *In Orbit, Probing the Natural World: Level III.* Intermediate Science Curriculum Study. Morristown, N.J.: Silver Burdett, 1972.

Name ....

Class ....

Date ....

Geocentric Theory vs. Heliocentric Theory for Venus

Worksheet

Position 1	Position 2	Position 3	Position 4
Phase	Phase	Phase	Phase ....
Position 5	**Position 6**	**Position 7**	**Position 8**
Phase	Phase	Phase	Phase ....

Figure 4. Sketches of Venus using the geocentric model.

Position 1	Position 2	Position 3	Position 4
Phase	Phase	Phase	Phase ....
Position 5	**Position 6**	**Position 7**	**Position 8**
Phase	Phase	Phase	Phase ....

Figure 5. Sketches of Venus using the heliocentric model.

Answer Key

Procedure

See the next page for the worksheet answer key.

Questions

1. The sketches of Venus as it actually appeared in the sky show that it exhibits all of the phases: crescent, quarter, gibbous, new, and full. When you view Venus using the geocentric model, all that you see is the crescent phase. Venus stays between the Earth and sun, so the lighted portion is always aimed away from the Earth. With the heliocentric model, Venus moves to Earth's side of the sun, and you see crescent phases. You see gibbous phases when Venus moves to the side of the sun that is opposite to the side the Earth is on, and more of its lighted portion is aimed toward the Earth. The fact that Venus goes through phases proves that the heliocentric theory for Venus is correct by demonstrating that the planet revolves around the sun.

2. The letter *A* represents sunrise, and *E* represents sunset.

3. Venus would be visible in the morning sky at positions 1, 2, and 3; in the evening sky, at positions 5, 6, and 7.

4. Venus appears smallest at positions 1, 7, and 8. When Venus appears full (position 8) or gibbous (positions 1 and 7), it is on the other side of its orbit from us. Because it is so far away, Venus appears smaller than it does in the crescent phases (positions 3 and 5), when it is on the side of its orbit that is closest to the Earth. Venus would appear largest to us at positions 3, 4, and 5. It would be difficult to see at position 4 because we would be looking into the sun.

5. One would think that Venus would appear the brightest during its full (position 8) or gibbous (positions 1 and 7) phases, when so much of its lighted part is aimed toward the Earth. Venus is actually brightest, however, when it is in its crescent phases (positions 3 and 5) because it is so much closer to the Earth at this time.

6. No, as seen from Venus, the Earth would never appear to be in a crescent or quarter phase. Earth is outside Venus's orbit, so Earth's lighted half, which is always aimed toward the sun, is also aimed toward Venus. To see a crescent Earth, Venus would have to move to the side of the Earth that is opposite to the side the sun is on, thus placing Venus outside Earth's orbit.

7. No, as seen from the Earth, Mars would never show crescent or quarter phases. Mars is to the Earth as the Earth is to Venus. The Earth's orbit is outside Venus's orbit, so the Earth can never be seen in a crescent or quarter phase from Venus. Mars' orbit is outside the Earth's orbit, so Mars can never be seen in a crescent or quarter phase from the Earth.

8. Any planet that is outside the viewer's orbit will always be seen as full or nearly full. Any planet that is inside the viewer's orbit will appear to go through all of the phases.

Answer Key

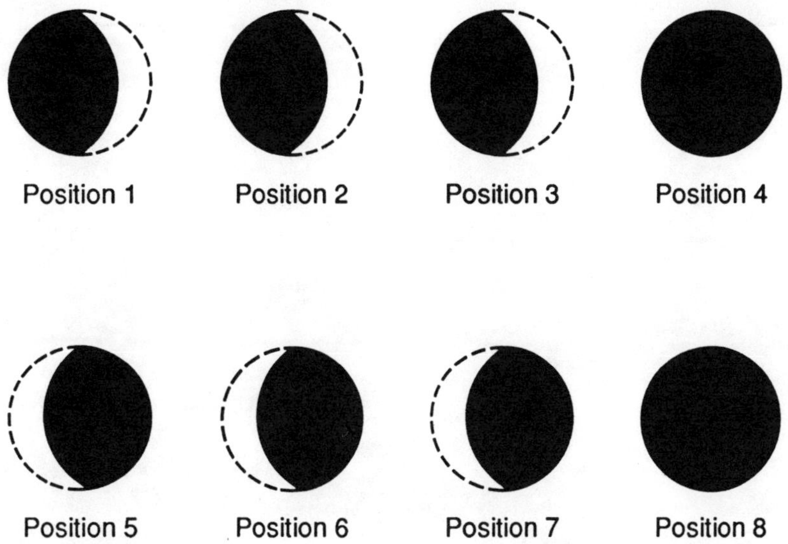

Figure 4. Sketches of Venus using the geocentric model. Positions 1, 2, 3, 5, 6, 7 show crescent phases. Positions 4 and 8 show new phases.

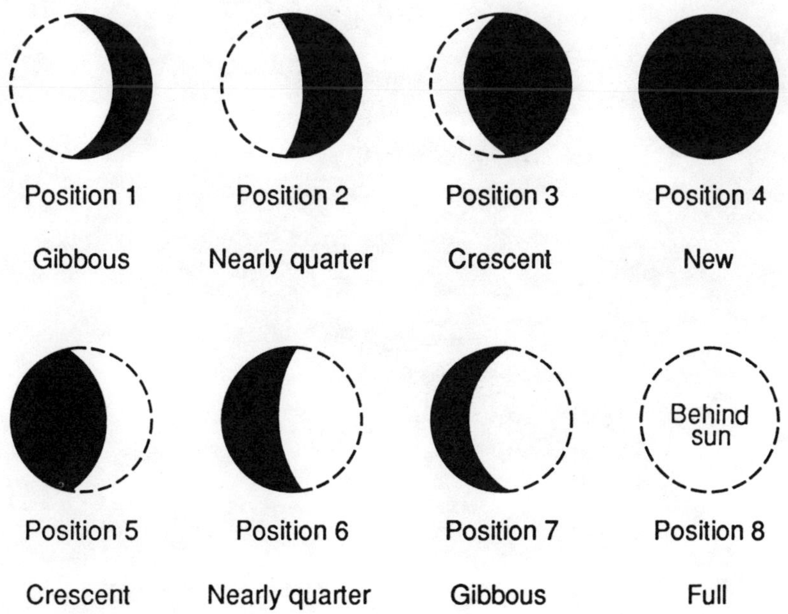

Figure 5. Sketches of Venus using the heliocentric model.

Instructions for Assembling Astronomy Materials

PAUL JOHNSTON
Emporia State University
Emporia, Kansas

The astronomy investigations, **Phases of the Moon, Solar and Lunar Eclipses,** and **Geocentric Theory vs. Heliocentric Theory for Venus,** require the materials described below.

CAUTION: *If students are involved in assembling the materials, be sure to emphasize safety procedures to be followed when using a hammer and drill.*

Time required

One hour

Materials

Note: English units are used instead of metric units because the items needed are generally available in English units.

Pegboard: 11 in. x 18 in.
2 strips of wood: 1 in. x 2 in. x 10 in.
4 nails: 1 in. long
Hammer
Drill
2 drill bits: 3/16 in. and 1/4 in.
2 dowels: 1/4-in. diameter, 3 in. long
3 polystyrene spheres: two with 1 1/2-in. diameters and one with a 1-in. diameter
Black paint: small amount for the spheres. Make sure the paint can be used on foam products.
Glue
3 golf tees
White tape: no wider than 1/4 in., 4 3/4 in. long
Block of wood: 2 in. x 2 in. x 3 in.

Procedure

1. Attach a 10-inch length of wood to each end of the pegboard to form legs (Figure 1). Each leg should be about three inches from either end. To avoid splitting the pegboard, nail the pegboard to the wood, not the wood to the pegboard. Use two nails for each strip.

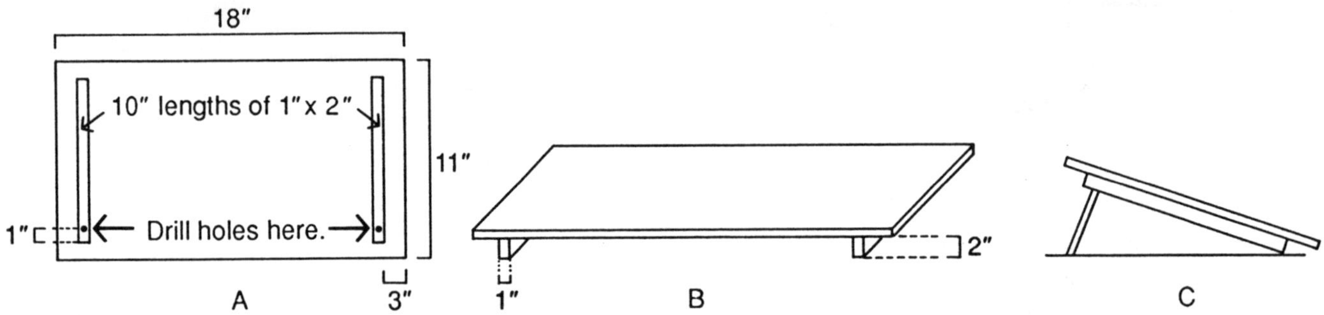

Figure 1. Pegboard with legs attached. (A) Bottom view. (B) Oblique view. (C) Side view with dowels attached.

2. Drill a 1/4-inch hole one inch from one end of each leg, as shown in Figure 1A. In **Solar and Lunar Eclipses**, a dowel will be placed in each hole to tilt the pegboard, as shown in Figure 1B.

3. Paint one 1 1/2-inch sphere half black; this represents the Earth. Leave the other 1 1/2-inch sphere white to represent the sun. Paint a one-inch sphere half black; this represents the moon or Venus, depending on which investigation is being done. Glue the spheres to golf tees (Figure 2).

4. Write the letters *A-H* equally spaced apart on a 4 3/4-inch piece of white tape. Moving counterclockwise, place the tape around the Earth sphere in the position of the equator. Place the letter *C* in the center of the white half (Figure 3).

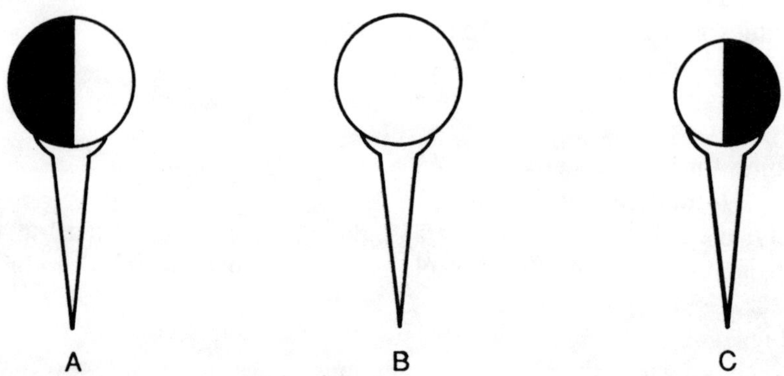

Figure 2. (A) 1 1/2-inch Earth sphere. (B) 1 1/2-inch sun sphere. (C) 1-inch moon or Venus sphere.

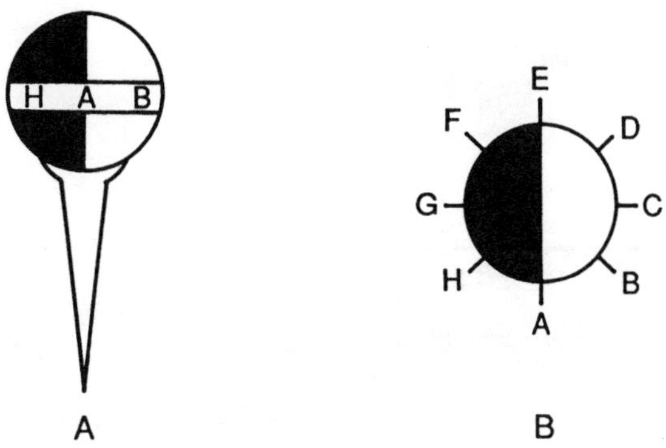

Figure 3. Earth sphere showing the position of the tape and letters.

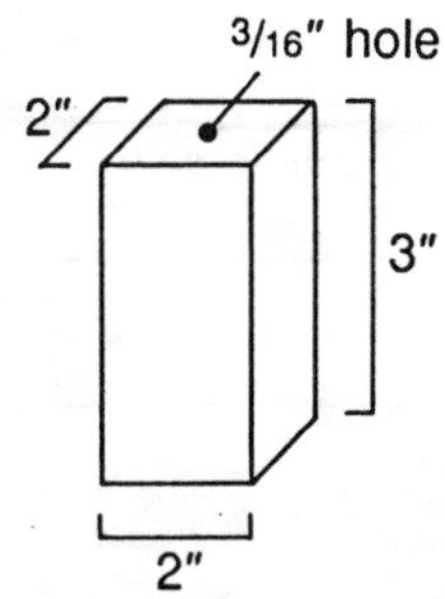

Figure 4. Block of wood showing hole on one end.

5. Drill a 3/16-inch hole in the center of one end of a 2 in. x 2 in. x 3 in. block of wood (Figure 4). The block will be used to support the sun golf tee in **Solar and Lunar Eclipses**.

6. Four overlays are included with the instructions. Each overlay consists of two pages to be placed on the pegboard for an investigation. The specific overlay and the investigation it is used with are listed below.

Overlay 1	**Phases of the Moon**
Overlay 2	**Solar and Lunar Eclipses**
Overlays 3 & 4	**Geocentric Theory vs. Heliocentric Theory for Venus**

The astronomy instructions can be modified to use polystyrene instead of pegboard, although the finished product will not be as sturdy or long-lasting. The materials to hammer and drill and to make the legs (two strips of wood) are not needed. Make the following substitutions:

Block of polystyrene for pegboard: about 2 in. thick
Block of polystyrene for block of wood
3 toothpicks for golf tees

If you want to assemble materials that will be even more durable than the pegboard unit, make the following substitutions:

Plywood for pegboard
Golf balls for polystyrene balls

Overlay 1 (left page)

Phases of the Moon

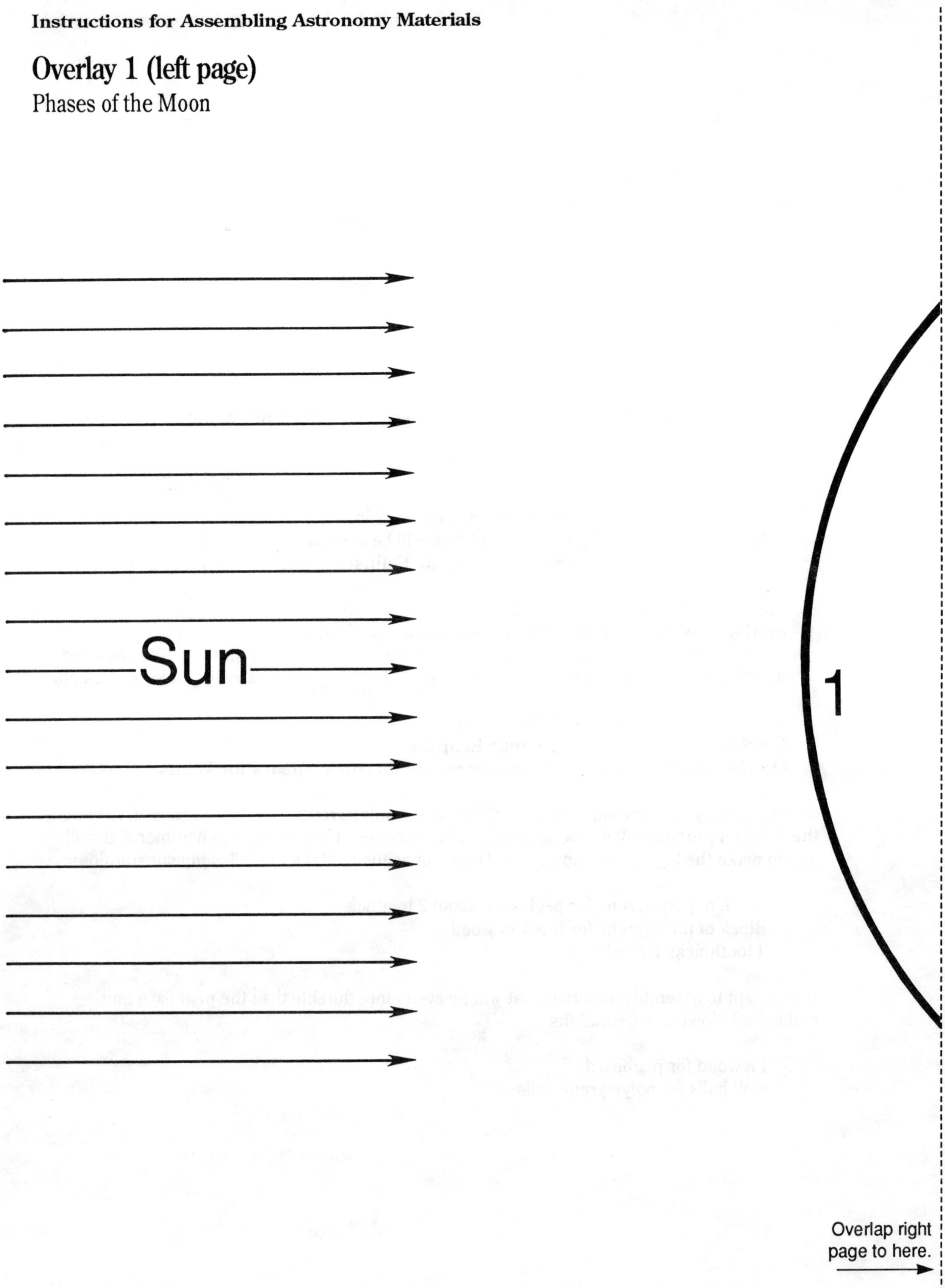

Overlay 1 (right page)

Phases of the Moon

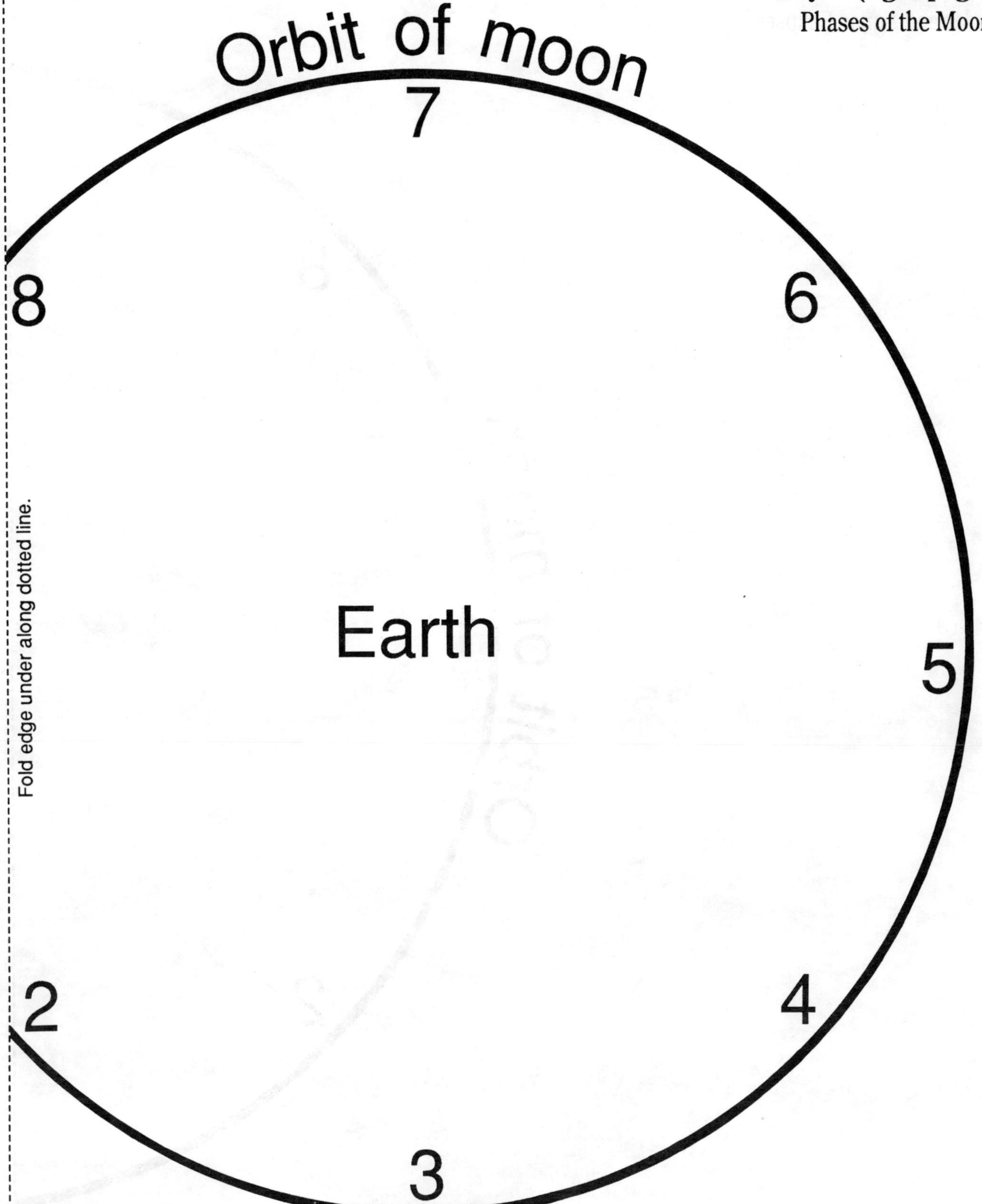

Fold edge under along dotted line.

Overlap onto left page.

Overlay 2 (left page)

Solar and Lunar Eclipses

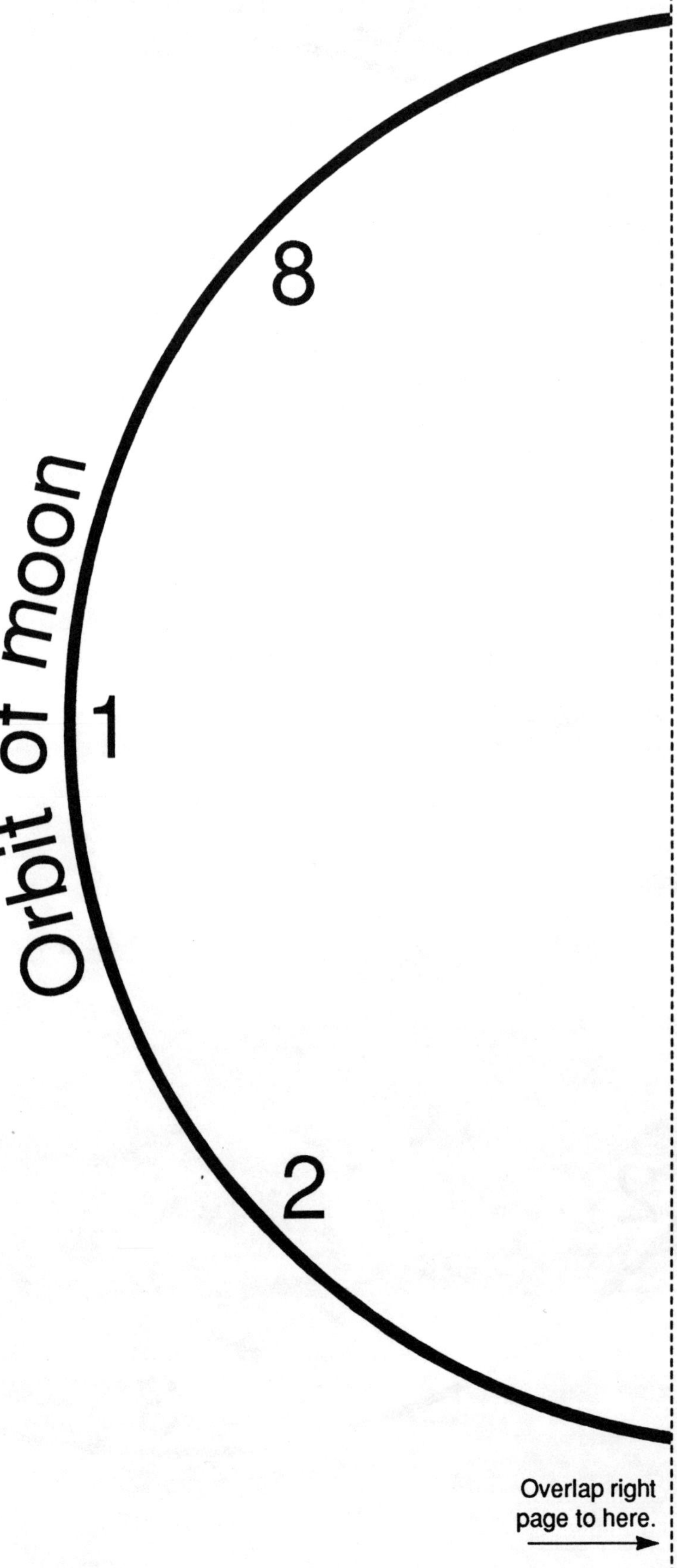

Overlay 2 (right page)

Solar and Lunar Eclipses

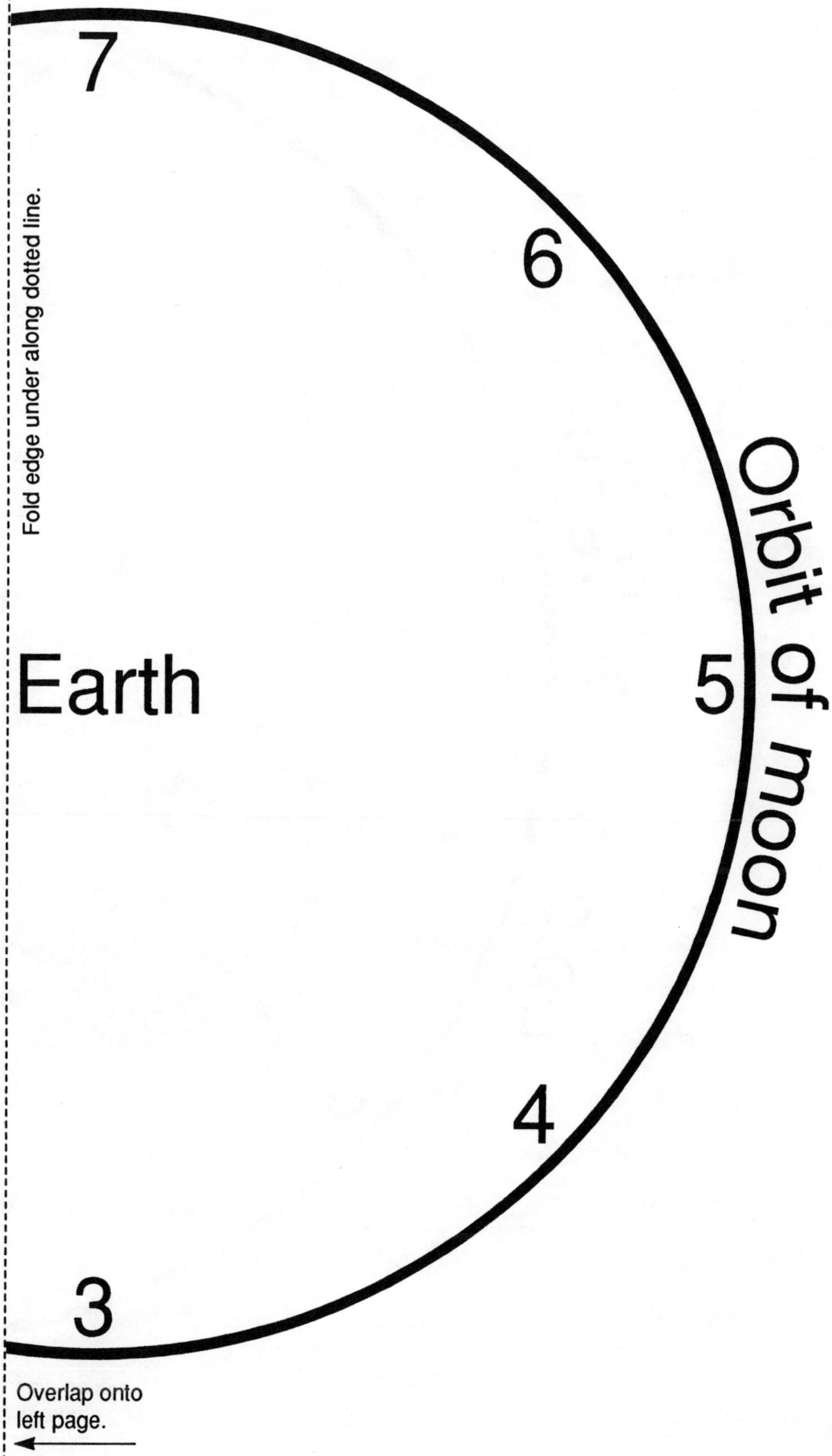

Overlay 3 (left page)

Geocentric Theory vs. Heliocentric Theory for Venus

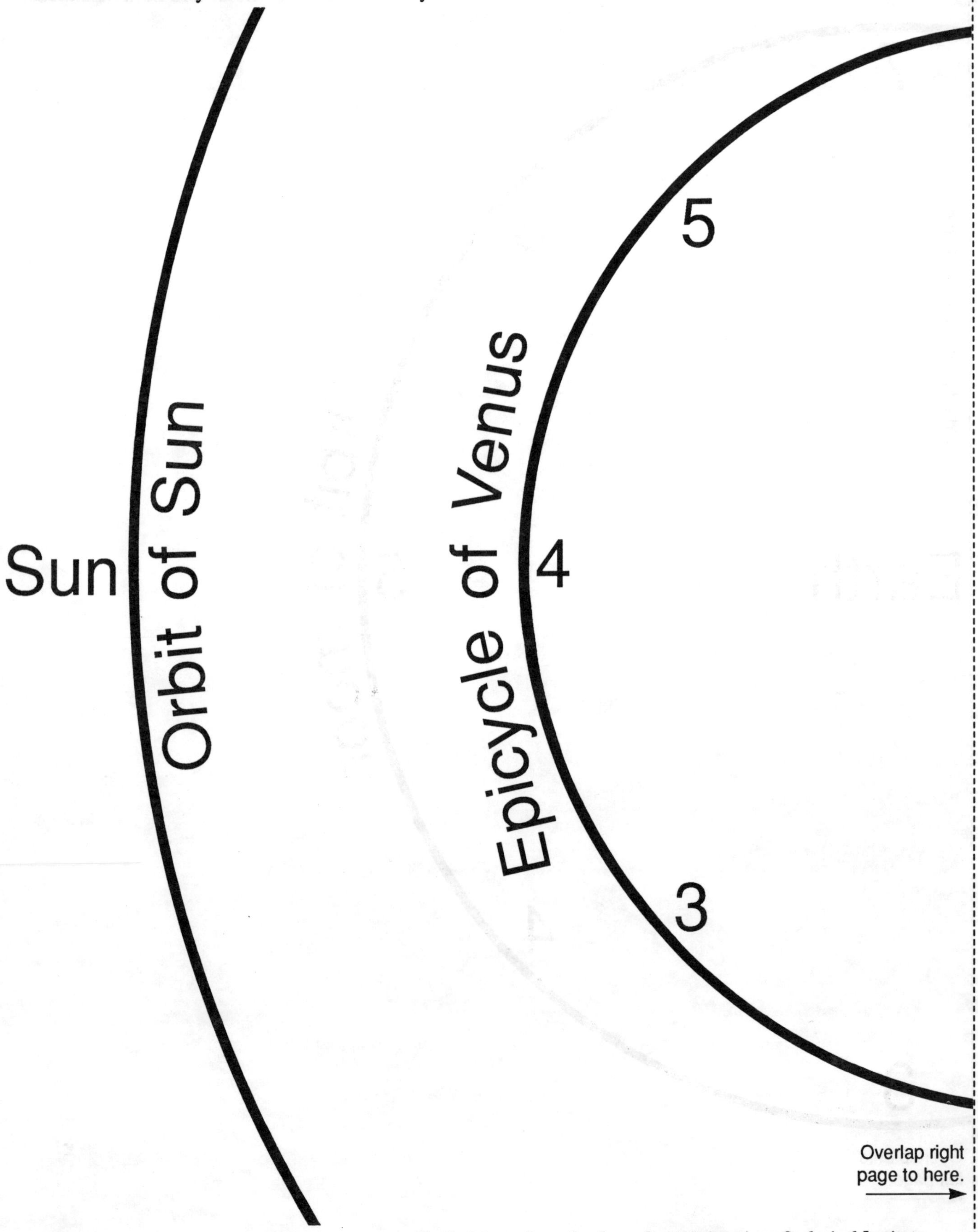

Overlay 3 (right page)

Geocentric Theory vs. Heliocentric Theory for Venus

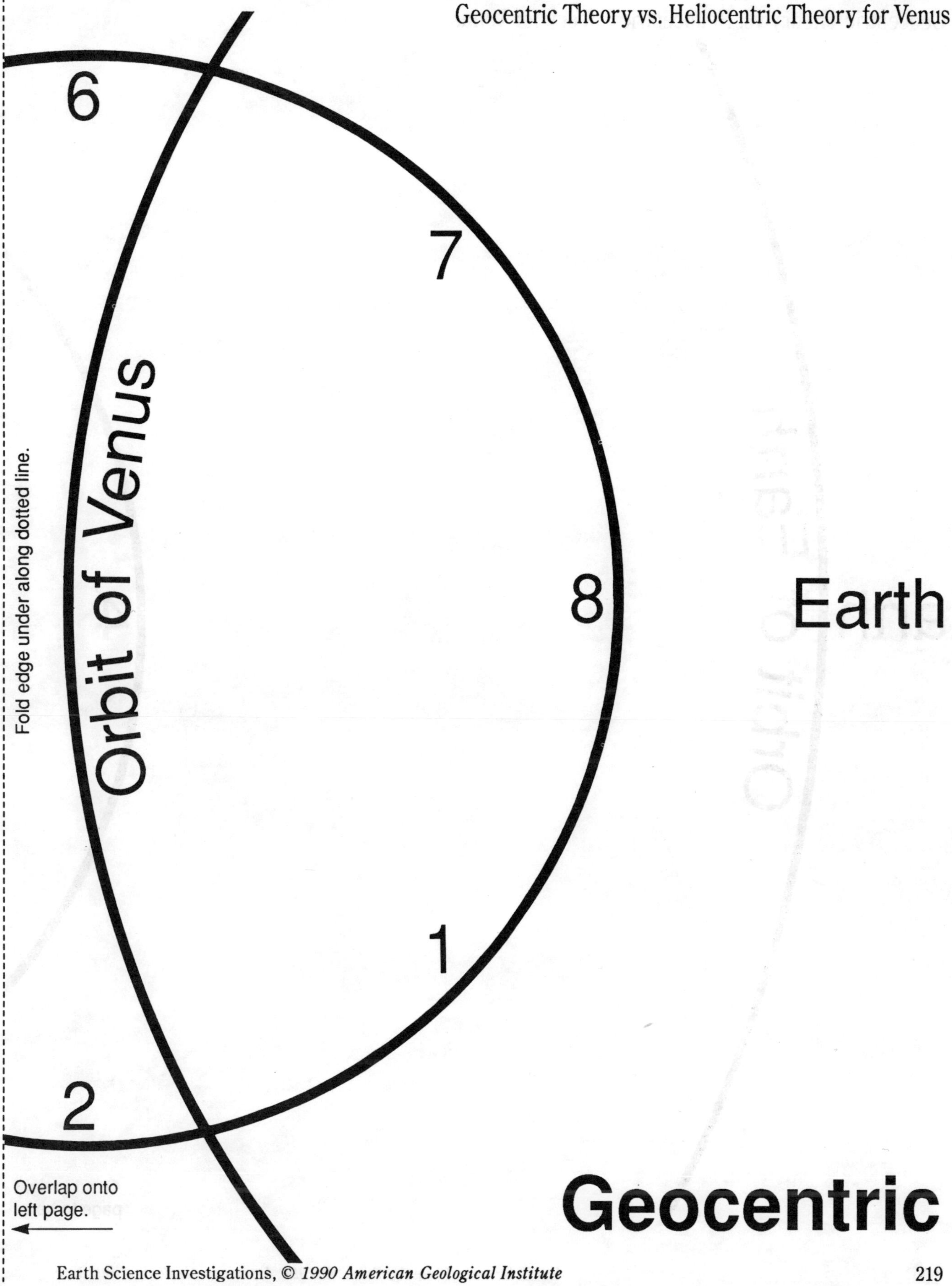

Overlay 4 (left page)

Geocentric Theory vs. Heliocentric Theory for Venus

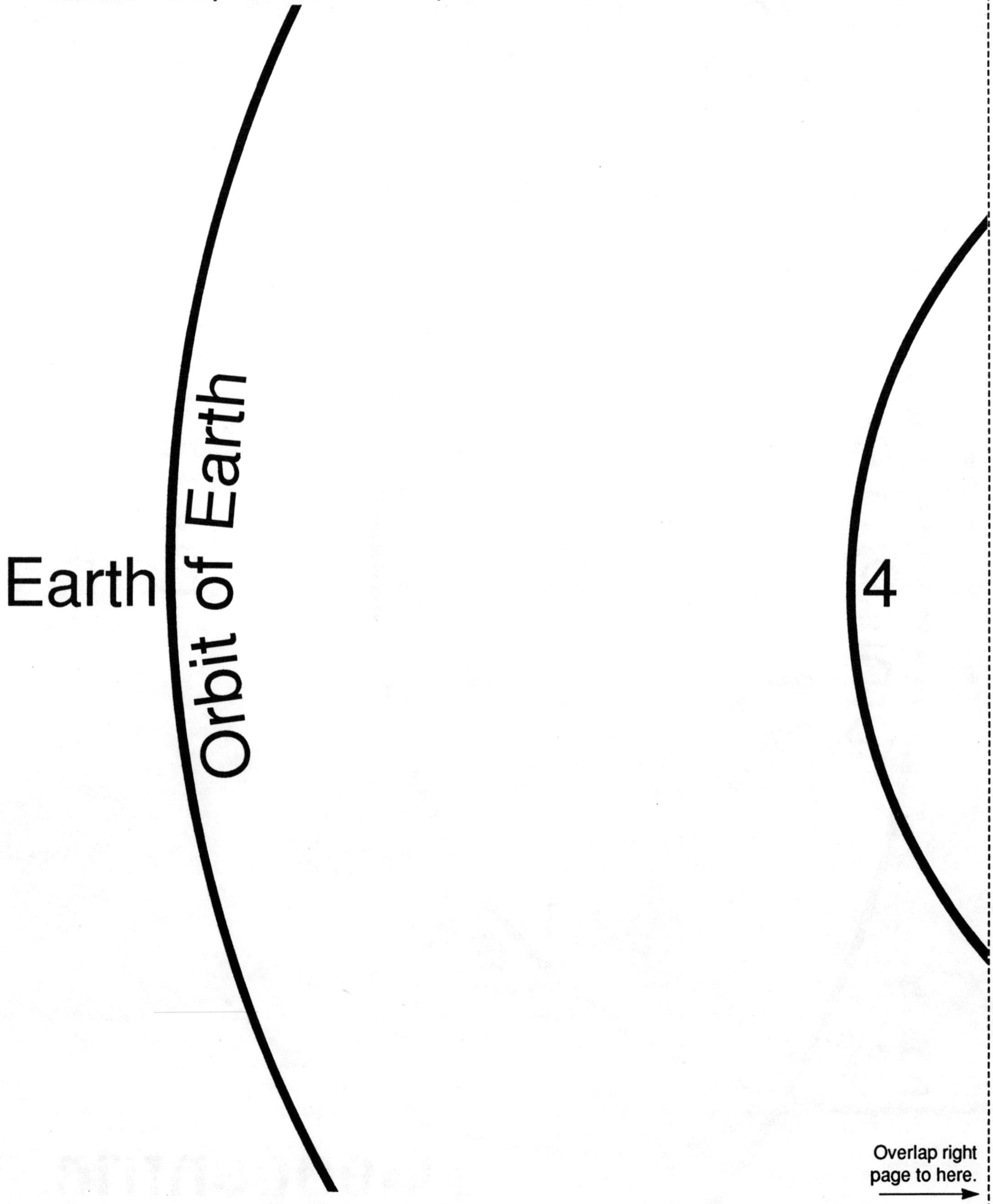

Overlay 4 (right page)

Geocentric Theory vs. Heliocentric Theory for Venus

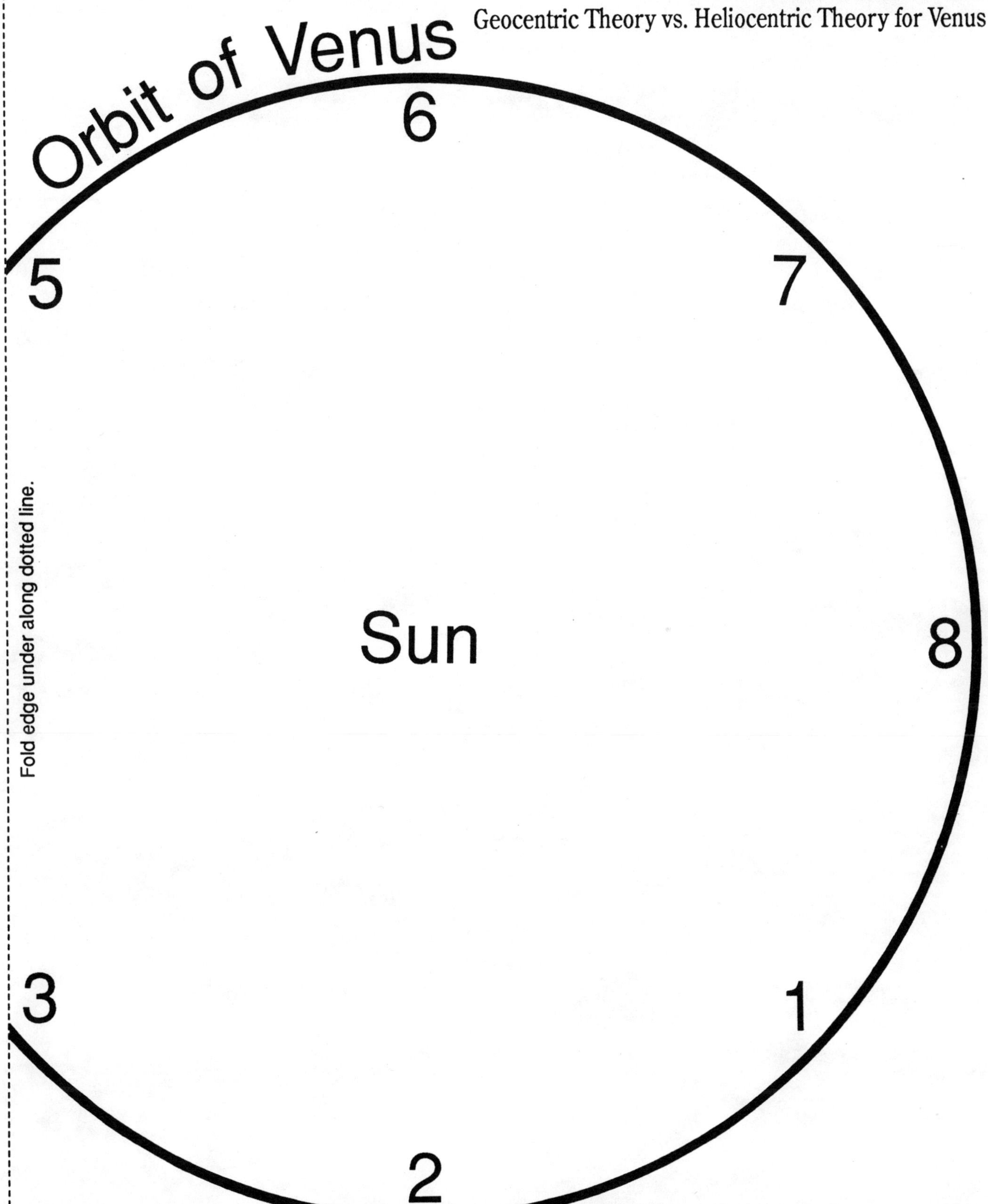

Analyzing North American Meteorite Impact Sites

JACOB M. HIRSCHFELT
Fredonia High School
Fredonia, New York

Objectives

To discover patterns in the meteorite impact history of North America and factors that may affect them.

To compare Earth impact sites with those on the moon and other bodies in our solar system.

Time required

Two class periods

Materials

per student

Map of the moon and/or Mercury showing distribution of craters. (See **Additional resources.**)
5 colored pencils: blue, red, green, black, and yellow
Computer graphing program (optional)

Background

Note: Students should know how to plot locations on a map using latitude and longitude and how to plot data on a graph.

Meteorite impacts on Earth and other celestial bodies give us an opportunity to learn more about our solar system. We can compile data on the rate and size of impacts through time, for example, and look for patterns and trends.

Several factors may alter impact sites and thereby make it more difficult to recognize and study them. Weathering, the natural process of breaking down rock materials, can disguise a site. Erosion may carry materials away from a site. Movements of the Earth's crust, volcanic activity, and human activity may also change or destroy evidence of meteorite sites.

Some of the factors above would have little or no effect on impact sites that are located on bodies other than the Earth. For example, the moon and Mercury do not show evidence of the type of large-scale crustal movements that occur on the Earth. These bodies also lack atmospheres; therefore craters on their surfaces are altered only by later meteorite impacts.

There is evidence to indicate that the moon and Mercury have had volcanic activity. It is possible that where few craters show today, lava has covered others. As you analyze and compare impact data, be sure to consider outside factors and their effects.

Procedure

Part I. Plotting the impact sites

Table 1 gives a symbol and color key to represent the size and age of a meteorite impact site. Table 2 lists 41 meteorite impact sites in North America. Using the Table 1 key, plot the 41 impact sites on the map of North America (Figure 1) on Worksheet 1. For example, you would plot a 23-km crater dated at 5 million years as a red dot. When plotting age, disregard the error estimates — the plus/minus values. The error estimates are included to show the degree of uncertainty surrounding the age determinations.

Table 1
Symbol and color key for meteorite impact sites.

Crater diameter (km)	Symbol	Age (million years)	Pencil color
≤25	•	<1	blue
26-50	+	1-9	red
≥51	*	10-99	green
		100-999	black
		≥1,000	yellow

Part II. Graphing crater size vs. crater age

Plot the size and age of the meteorite craters that are listed in Table 2 on the graph (Figure 2) on Worksheet 2, or use a computer graphing program to plot the data. Ignore the error estimates when plotting age.

Questions

1. Describe the overall distribution of craters on your map.
2. Compare the distribution of craters on a map of the moon and/or Mercury with that on your map of North American impact sites. Describe the similarities and differences you see. What factors might explain each?
3. How might the fact that much of northern Canada is relatively uninhabited affect the number of reported craters there?
4. Describe the distribution of craters on your graph.
5. How could the size of a crater affect how long it could last?
6. What other factors could influence how long a crater lasts on Earth?
7. No craters have been found with ages between 120 to 200 million years and 600 to 1,850 million years. What reasons can you suggest to explain these gaps? Note the amount of error in some of the age estimates.
8. Calculate the average number of impacts per 10 million years with and without the Sudbury impact (1,850 million years old). Disregard the error estimates in your calculations. Is North America due for another large impact? Explain your answer.

Additional resources

Hartmann, William K. "Cratering in the Solar System." *Scientific American* 236, no. 1 (January 1977): 84-99.

Murray, Bruce C. "Mercury." *Scientific American* 233, no. 3 (September 1975): 58-68.

Weaver, Kenneth F. "The Moon — Man's First Goal in Space." *National Geographic* 135, no. 2 (February 1969): 206-229.

_________. "Have We Solved the Mysteries of the Moon?" *National Geographic* 144, no. 3 (September 1973): 308-325.

_________. "Mariner Unveils Venus and Mercury." *National Geographic* 147, no. 6 (June 1975): 858-869.

_________. "Meteorites — Invaders from Space." *National Geographic* 170, no. 3 (September 1986): 390-418.

Wood, John A. "The Moon." *Scientific American* 233, no. 3 (September 1975): 92-102.

Maps of the moon

The Earth's Moon (1976).
National Geographic Society
Educational Services
Department 90
Washington, D.C. 20036
(1-800-368-2728)

Exploring the Moon.
Sky Publishing Co.
49 Bay State Rd.
Cambridge, Mass. 02138
(617-864-7360)

Map of Mercury

Mercury Reference Mosaic, Number I-903 (1974).
U.S. Geological Survey
For ordering information, contact any of the U.S. Geological Survey Earth Science Information Centers listed on the inside front cover.

Table 2
North American meteorite impact sites.

	Latitude	Longitude	Crater diameter (km)	Age (millions of years)
1. Barringer, Meteor Crater, Arizona	35°N	111°W	1.2	0.025
2. Bee Bluff, Texas	29°N	100°W	2.4	<40
3. Brent, Ontario	46°N	78°W	3.8	450 ±30
4. Carswell, Saskatchewan	58°N	110°W	37.0	117 ±8
5. Charlevoix, Quebec	48°N	70°W	46.0	360 ±25
6. Clearwater Lake East, Quebec	56°N	74°W	22.0	290 ±20
7. Clearwater Lake West, Quebec	56°N	75°W	32.0	290 ±20
8. Crooked Creek, Missouri	38°N	91°W	5.6	320 ±80
9. Decaturville, Missouri	38°N	93°W	6.0	<300
10. Deep Bay, Saskatchewan	56°N	103°W	12.0	100 ±50
11. Eagle Butte, Alberta	50°N	110°W	10.0	<65
12. Flynn Creek, Tennessee	36°N	86°W	3.8	360 ±20
13. Glover Bluff, Wisconsin	44°N	90°W	6.0	<500
14. Gow Lake, Saskatchewan	56°N	104°W	5.0	<250
15. Haughton, Northwest Territory	75°N	90°W	20.0	21.5 ±1.2
16. Haviland, Kansas	38°N	99°W	0.011	<0.001
17. Holleford, Ontario	44°N	77°W	2.0	550 ±100
18. Ile Rouleau, Quebec	51°N	74°W	4.0	<300
19. Kentland, Indiana	41°N	87°W	13.0	<300
20. Lac Couture, Quebec	60°N	75°W	8.0	425 ±25
21. Lac la Moinerie, Quebec	57°N	67°W	8.0	400 ±50
22. Lake St. Martin, Manitoba	52°N	99°W	23.0	225 ±40
23. Lake Wanapitei, Ontario	47°N	81°W	8.5	37 ±2
24. Manicouagan, Quebec	51°N	69°W	100.0	210 ±4
25. Manson, Iowa	43°N	95°W	32.0	61 ±9
26. Marquez Dome, Texas	37°N	96°W	14.0	58 ±8
27. Middlesboro, Kentucky	37°N	84°W	6.0	<300
28. Mistastin Lake, Labrador	56°N	63°W	28.0	38 ±4
29. Montagnais Crater, Nova Scotia	43°N	64°W	45.0 ?	50.5 ±0.8
30. New Quebec Crater, Quebec	61°N	74°W	3.2	1.0 ±0.3
31. Nicholson Lake, Northwest Territory	63°N	103°W	12.5	<400
32. Odessa, Texas	32°N	102°W	0.17	0.03
33. Pilot Lake, Northwest Territory	60°N	111°W	6.0	440 ±2
34. Red Wing Creek, North Dakota	48°N	104°W	9.0	200
35. Serpent Mound, Ohio	39°N	83°W	6.4	<320
36. Sierra Madera, Texas	31°N	103°W	13.0	<100
37. Slate Islands, Ontario	49°N	87°W	30.0	<350
38. Steen River, Alberta	60°N	118°W	25.0	95 ±7
39. Sudbury, Ontario	47°N	81°W	140.0	1,850 ±150
40. Wells Creek, Tennessee	36°N	88°W	14.0	200 ±100
41. West Hawk Lake, Manitoba	50°N	95°W	2.7	100 ±50

Source: *Observer's Handbook - 1990.* (Toronto: Royal Astronomical Society of Canada). Published annually.

Note: The 41 craters above represent about one-third of the craters discovered worldwide. There are at least 20 other suspected but unconfirmed North American impact sites. Most crater ages have been determined by radiometric dating techniques.

Name..
Class...
Date..

Analyzing North American Meteorite Impact Sites

Worksheet 1

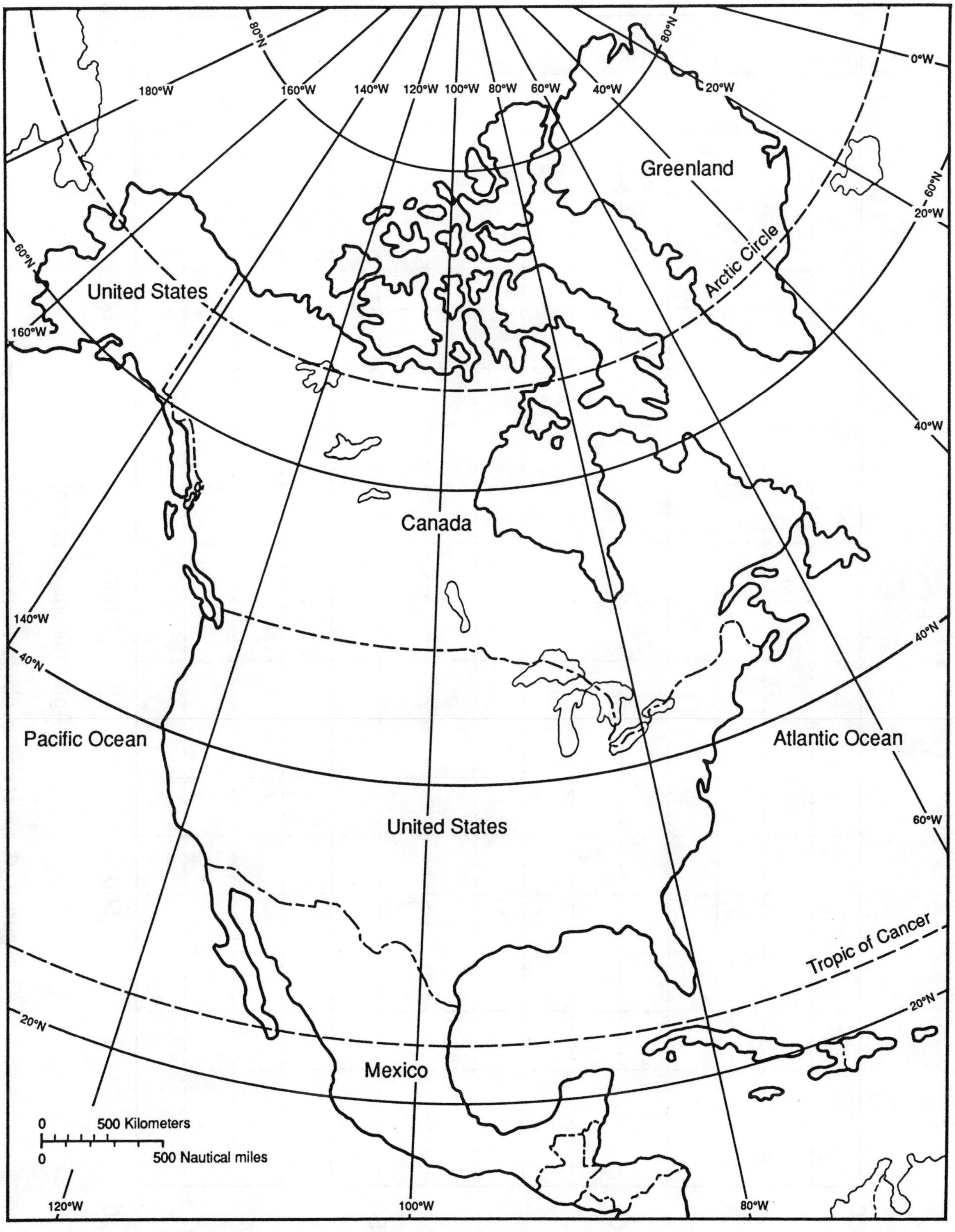

Figure 1. Map of North America.

Name ....

Class ....

Date ....

Worksheet 2

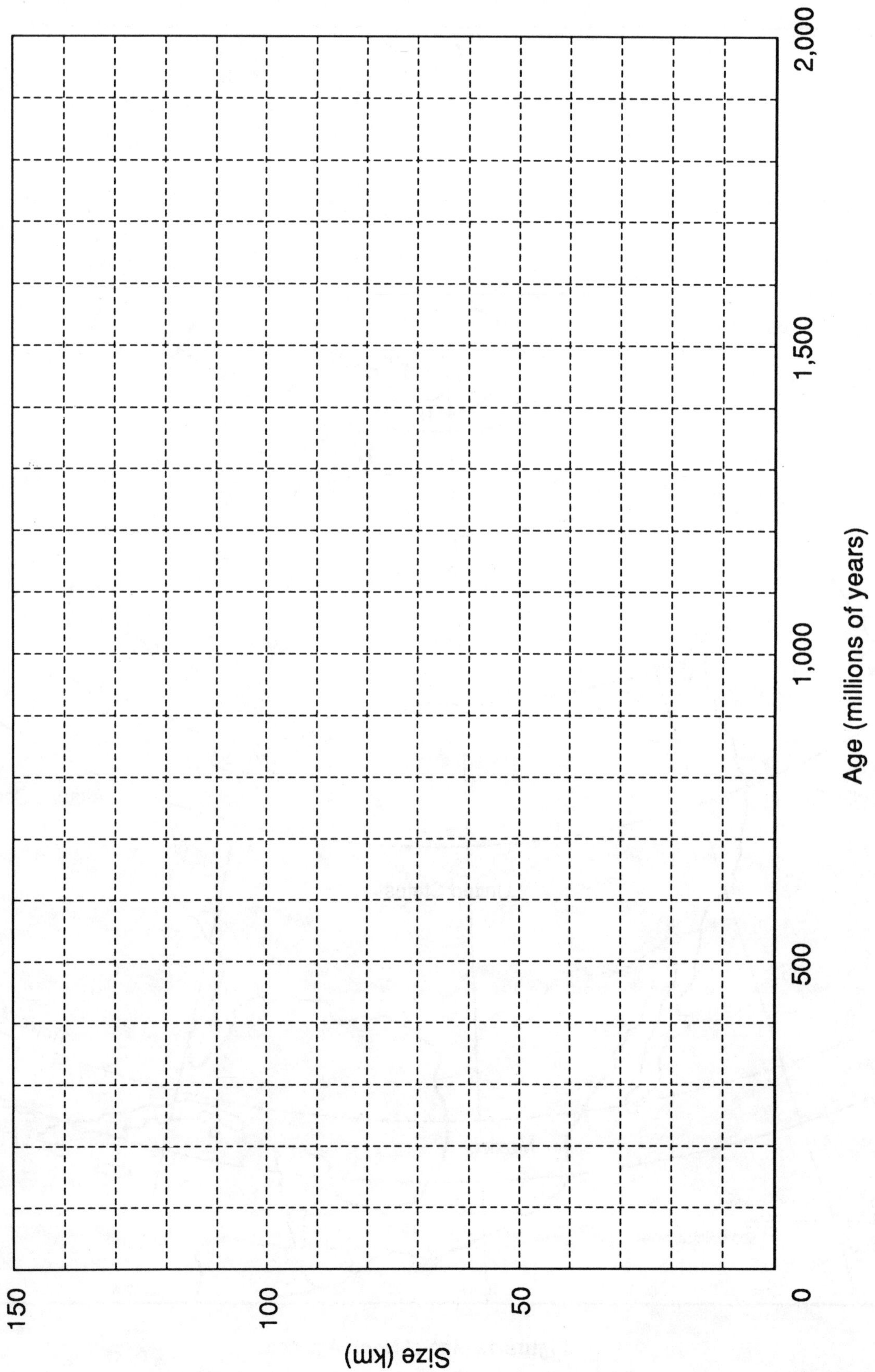

Figure 2. Graph of size vs. age for North American impact craters.

Answer Key

Procedure

See the next pages for the worksheet answer keys.

Questions

1. The map shows that most of the impacts occurred in the central region of North America. No impacts are recorded for Mexico, and only one known impact occurred in the western part of the United States.

2. The moon and Mercury are heavily cratered. Some craters overlap. Craters are within craters. Since neither the moon nor Mercury has an atmosphere, the craters have not been weathered by water or wind. The moon and Mercury do not show evidence of large-scale crustal movements. Impact events on the Earth may have been as numerous as on the moon (they are very close together in the solar system). Most of the craters on the Earth, however, have been altered or destroyed by weathering or crustal movements.

3. The probability of locating a crater would increase if more people inhabited a region. They would probably build more roads and explore a greater portion of the area. As more wells were drilled in the area and data from the subsurface were analyzed, the chances of locating buried craters would increase.

4. Two craters are very large, 100 km or more in diameter; the rest are less than 50 km in diameter. The largest crater, the Sudbury impact, is also the oldest. It is nearly 2,000 million years old, while the rest are 550 million years old, or younger. Most of the identified craters are relatively young and small.

5. The largest crater in North America is about 1,850 million years old. The larger a crater is, the greater the probability that erosion will not completely remove it and that some evidence of the crater will be found.

6. A crater's location may determine how long it will last on Earth. A crater in a humid environment may weather faster than one in a dry environment. Glacial or volcanic activity may cover evidence of an impact. Impacts in the oceans would be difficult to locate.

7. Some possible reasons for why no craters between the ages of 120 to 200 million years and 600 to 1,850 million years have been found are listed below.
   a. The craters have weathered away because they were small.
   b. Movements of the Earth's crust may have altered the regions in which impacts occurred, destroying evidence that a site existed.
   c. Craters of these time periods may yet be found. Some remote areas still need to be explored more extensively.
   d. The dating of some craters has a large margin of error (sometimes as great as 50 million or more years). Better dating of these impact sites may show that some of the craters on the list belong to one of these time periods.
   e. At least 20 suspected impact sites in Canada and the United States are not listed. Some of these may fall into one of the two age ranges.

8. (41 craters/1,850 million years) x 10 = 0.22 craters per 10 million years

   (40 craters/550 million years) x 10 = 0.73 craters per 10 million years

   No, considering the impact rate calculated above, North America is not due for another large impact. The youngest known crater larger than one km in diameter in North America is 25,000 years old (Meteor Crater). From the data, there appears to be no pattern to the timing of an impact. One could occur any day, but the likelihood of a major impact hitting North America soon is small.

Answer Key

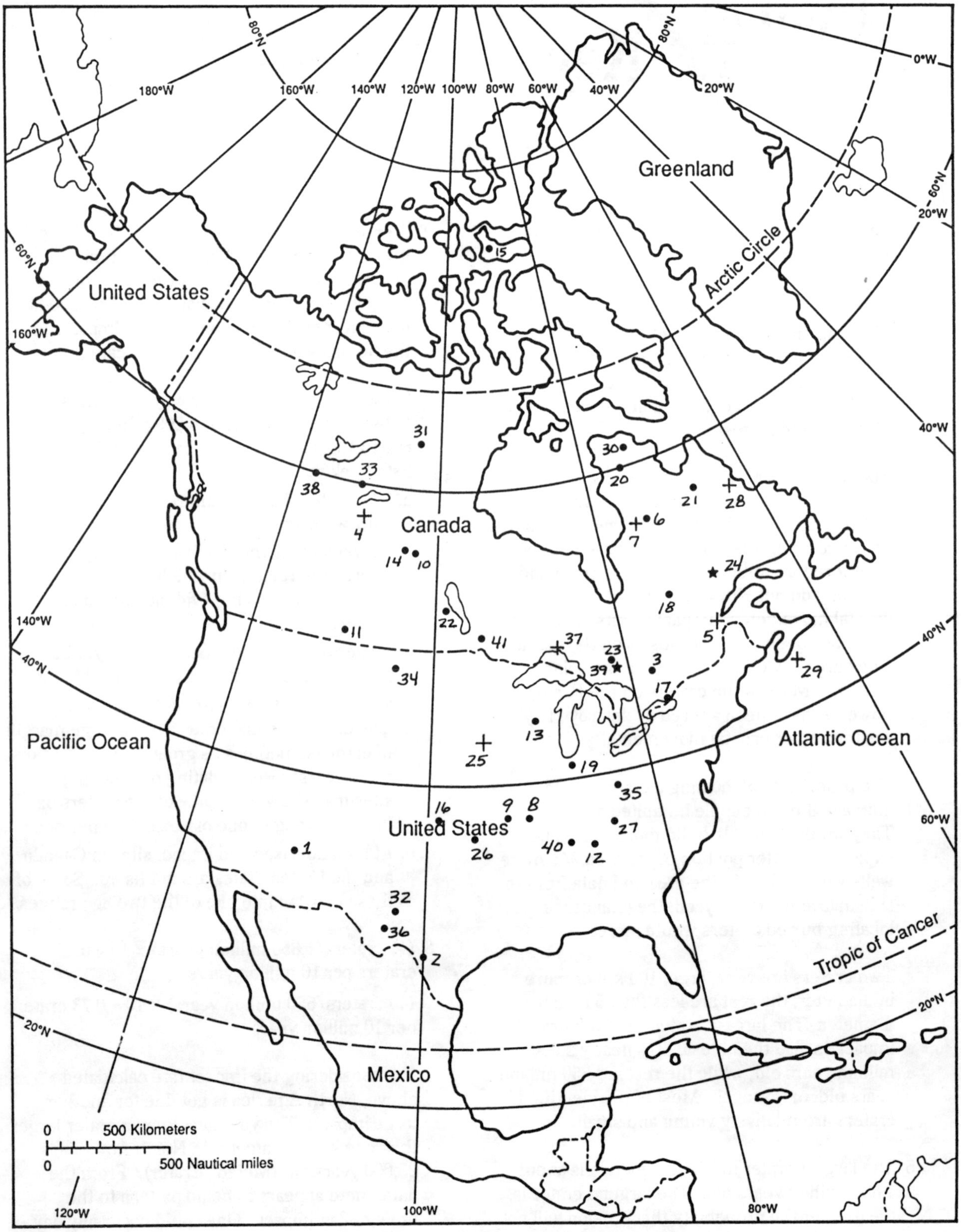

Figure 1. Map of North America.

Answer ✓ Key

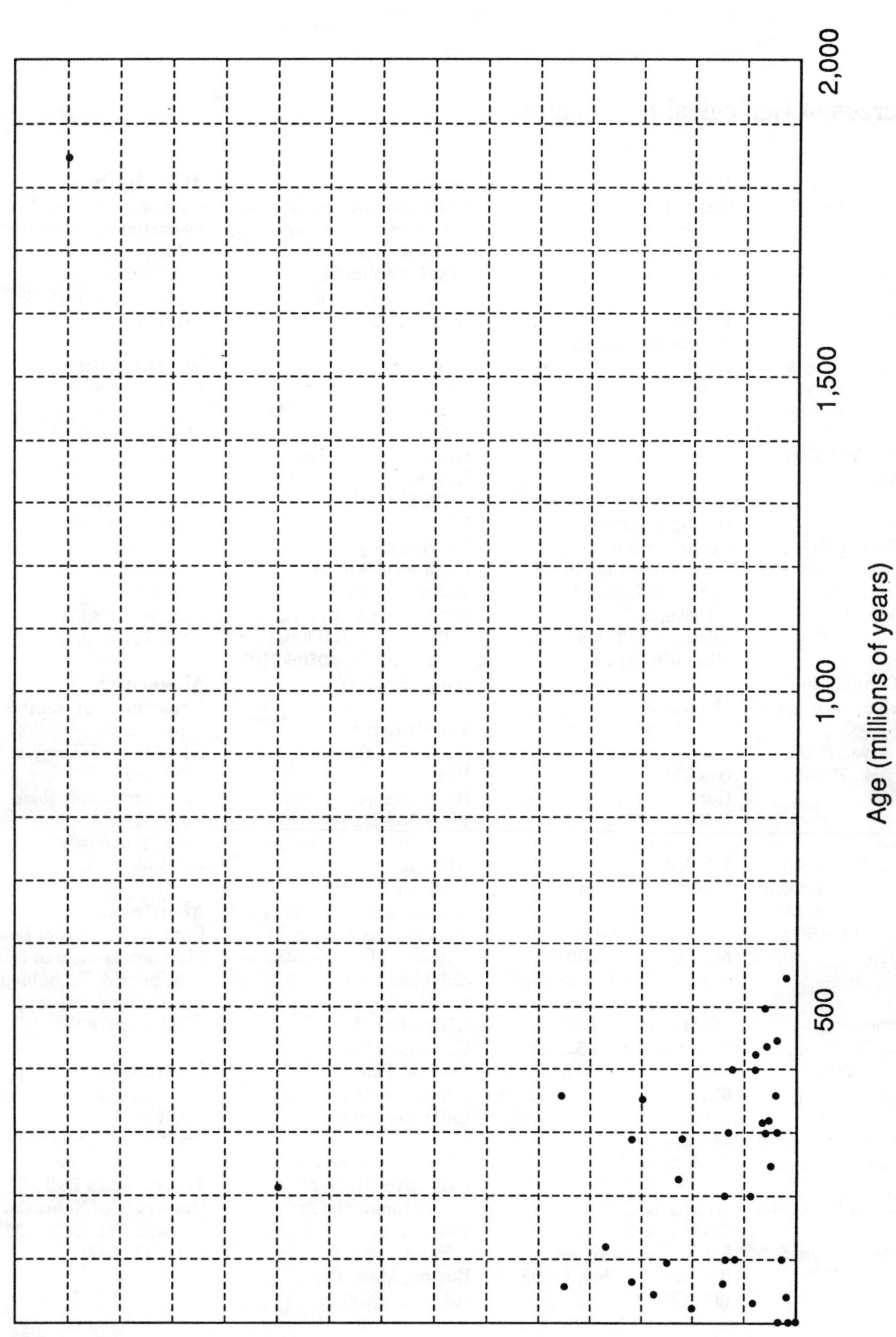

Figure 2. Graph of size vs. age for North American impact craters. In two instances, the locations of two impact sites plot so closely on the graph as to be represented by one point: sites 9 and 27; sites 16 and 32.

State Sources of Geological Information

Alabama
Geological Survey of Alabama
Box O
Tuscaloosa, Ala. 35486-9780
(205) 349-2852

Alaska
Division of Geological & Geophysical Surveys
794 University Ave., Suite 200
Fairbanks, Alaska 99709
(907) 479-7625

Arizona
Geological Survey Branch
845 N. Park Ave., Suite 100
Tucson, Ariz. 85719
(602) 621-7906

Arkansas
Geological Commission
Vardelle Parham Geology Center
3815 W. Roosevelt Road
Little Rock, Ark. 72204
(501) 371-1488

California
Department of Conservation
Division of Mines & Geology
1416 Ninth St., Room 1341
Sacramento, Calif. 95814
(916) 445-1923

Colorado
Geological Survey
1313 Sherman St., Room 715
Denver, Colo. 80203
(303) 866-2611

Connecticut
Natural Resources Center
Department of Environmental Protection
State Office Building
165 Capitol Ave., Room 553
Hartford, Conn. 06106
(203) 566-3540

Delaware
Geological Survey
University of Delaware
Newark, Del. 19716
(302) 451-2833

Florida
Bureau of Geology
Department of Natural Resources
903 W. Tennessee St.
Tallahassee, Fla. 32304-7795
(904) 488-4191

Georgia
Geologic Survey
Department of Natural Resources, Room 400
19 Martin Luther King, Jr., Drive, S.W.
Atlanta, Ga. 30334
(404) 656-3214

Hawaii
Division of Water & Land Development
Box 373
Honolulu, Hawaii 96809
(808) 548-7533

Idaho
Geological Survey
University of Idaho
Morrill Hall, Room 332
Moscow, Idaho 83843
(208) 885-7991 or 885-6195

Illinois
State Geological Survey
Natural Resources Building
615 E. Peabody Drive, Room 121
Champaign, Ill. 61820
(217) 333-5111 or 333-4747

Indiana
Geological Survey
611 N. Walnut Grove
Bloomington, Ind. 47405
(812) 335-2862

Iowa
Geological Survey Bureau
Department of Natural Resources
123 N. Capitol St.
Iowa City, Iowa 52242
(319) 335-1575

Kansas
Geological Survey
1930 Constant Ave., West Campus
University of Kansas
Lawrence, Kan. 66046
(913) 864-3965

Kentucky
Geological Survey
University of Kentucky
228 Mining & Mineral Resources Building
Lexington, Ky. 40506-0107
(606) 257-5500

Louisiana
Geological Survey
Box G
Baton Rouge, La. 70893
(504) 388-5320

Maine
Geological Survey
Department of Conservation
State House, Station 22
Augusta, Maine 04333
(207) 289-2801

Maryland
Geological Survey
2300 St. Paul St.
Baltimore, Md. 21218
(301) 554-5503

Massachusetts
Executive Office of Environmental Affairs
100 Cambridge St., 20th Floor
Boston, Mass. 02202
(617) 727-9800

Michigan
Geological Survey Division
Department of Natural Resources
Box 30028
Lansing, Mich. 48909
(517) 334-6923

Minnesota
Geological Survey
2642 University Ave.
St. Paul, Minn. 55114-1057
(612) 627-4780

Mississippi
Bureau of Geology
Department of Natural Resources
Box 5348
Jackson, Miss. 39216
(601) 354-6228

Missouri
Department of Natural Resources
Division of Geology & Land Survey
111 Fairgrounds Road
Box 250
Rolla, Mo. 65401
(314) 364-1752

Montana
Bureau of Mines & Geology
Montana College of Mineral Science & Technology
Butte, Mont. 59701
(406) 496-4180

Nebraska
Conservation & Survey Division
Institute of Agriculture & Natural Resources
113 Nebraska Hall
University of Nebraska
Lincoln, Neb. 68588-0517
(402) 472-3471